MW00966913

PIXIES,
SIX-PACKS,
AND SUPERMEN

Mike Olszewski and Richard Berg

With Carlo Wolff

Black Squirrel Books ❧ Kent, Ohio

BLACK SQUIRREL BOOKS™

In 1961 Kent State University's grounds superintendent imported ten black squirrels from Ontario, Canada, to Kent, Ohio. In the ensuing decades the squirrels have
thrived and multiplied, becoming a familiar sight on the lawns and in the trees of
the Kent State campus. These frisky, industrious creatures are now a Kent State
icon, the mascot of KSU's annual Black Squirrel Festival, and the inspiration for
Black Squirrel Books™, a trade imprint of The Kent State University Press.

WWW.KENTSTATEUNIVERSITYPRESS.COM

© 2011 by The Kent State University Press, Kent, Ohio 44242
ALL RIGHTS RESERVED
Library of Congress Catalog Card Number 2011016807
ISBN 978-1-60635-099-7
Manufactured in the United States of America

LIBRARY OF CONGRESS CATALOGING-IN-PUBLICATION DATA
Olszewski, Mike, 1953–
WIXY 1260 : pixies, six-packs, and supermen /
Mike Olszewski and Richard Berg ; with Carlo Wolff.
p.   cm.
Includes bibliographical references and index. ∞
ISBN 978-1-60635-099-7 (pbk. : alk. paper)
1. WIXY (Radio station : Cleveland, Ohio)—History. 2. Radio stations—Ohio—
Cleveland—History. 3. Rock music—History and criticism. I. Berg, Richard. II.
Wolff, Carlo. III. Title.
HE8698.O54 2011
791.4409771'32—dc23
2011016807

British Library Cataloging-in-Publication data are available.

15 14 13 12 11   5 4 3 2 1

This book is dedicated to the
most important people in our lives,

JANICE OLSZEWSKI

and

ANNE BERG

CONTENTS

FOREWORD

In the radio heavens there are but a few truly great radio stations that shine as stars and reflect the Top 40 era of the 1960s and early 1970s with brilliance. Some of those great radio stations were KLIF in Dallas, KAAY in Little Rock, WIL in Saint Louis, KILT in Houston, KHOW in Denver, WABC in New York, WHB in Kansas City, KFRC in San Francisco, WDRC in Hartford, WAKY in Louisville, WHBQ in Memphis, CKLW in Windsor Ontario, KXOL in Fort Worth, KHJ in Los Angeles, WXYZ in Detroit, WKBW in Buffalo, WKY in Oklahoma City, KOIL in Omaha, WOKY in Milwaukee, WTIX in New Orleans, WNOE in Baton Rouge, KIMN in Denver, KQV in Pittsburgh, KCBQ in San Diego, KJR in Seattle, WFUN in Miami, KONO in San Antonio, and many more. Among them are the call letters of a small-powered radio station that achieved monumental status . . . WIXY in Cleveland. There will never be another radio station more perfectly executed and promoted as WIXY 1260. It was one of a kind and working there for several years was sometimes a matter of hanging on and letting the magic happen. It's easy to lose your objectivity when you're so closely involved with the subject about which you're writing. Mike and Richard approached the history of WIXY 1260 with open and curious minds, backed up with thorough research. In 1968 I took a DJ position at WKYC, which was an NBC owned and operated radio station located across town from WIXY. WKYC tried to dethrone WIXY as the ratings leader in Cleveland, implementing the same format but not as well. WIXY's "Three in a Row" at the top of every hour, great air talent, and excellent promotions were too tough to beat. After a year WKYC abandoned the project and changed formats to easy listening. That is the day I changed radio stations thanks to Norm Wain and joined WIXY 1260, the Cleveland ratings leader as afternoon DJ and music director. I remained there for several wonderful years. As I mentioned earlier, it is sometimes difficult to see the forest for the trees but this book gives us readers a clear and objective look at one of the great radio stations in America in the late 1960s and early 1970s. I had the honor of being Disc Jockey, Music Director, and eventually Program Director over the course of several years and always knew I was blessed to be a part of this magical radio station. The people, promotions, and the politics of it all unfold in the pages you are about to read. Hats off to Mike and Richard, who have labored for several years to put it all into perspective and catalog it for history. Anyone who grew up listening to the legendary WIXY will be rewarded with a voyage through time, exploring a radio station that can never be duplicated.

Chuck Dunaway (retired)
Texas Radio Hall of Fame 2002
Radio/Television Broadcasters Hall of Fame 2003 of Ohio

ACKNOWLEDGMENTS

Our thanks go out to some very special people. This book could not have seen print without the tireless efforts (and patience) of Carlo Wolff. Carlo is a renowned pop culture historian in his own right and a truly gifted writer. We don't think he knew what he was getting into when he agreed to edit this book, but we cannot thank him enough for his hard work and insight.

We also thank Bill Barrow and the Cleveland Press Collection at Cleveland State University, Ray Berg, Fred Bourjaily, Brad Funk, Eric Funk, Rick Funk, Nicole Fiorilli, Tracy Fiorilli, Andrew Heinzman, Peter Heinzman, everyone at the Kent State University Press, Bianca Kontra, Angelina Leas, Connie Little, Dave Little, Cate Misciagna, Cole Misciagna, Jenny Misciagna, Tony Misciagna, Landen Phillips, Matt Phillips, and Theresa Phillips.

PROLOGUE

More than any other media, radio reminds members of the generation who came of age in the 1960s and 1970s of youth—whether as a transistor hidden under the pillow or blasting out of a speaker in a first car, radio is the great equalizer. It's the one topic at the high school reunion that everyone can relate to, and the older we get, the more we cherish the memories of that little box we held against our ears. No matter which side of town we were from or how much money our parents had, we were all on equal ground when we listened to our favorite station.

Every town has its pet radio stations and deejays, but the Cleveland market attracted superstars. It had a long-standing tradition of honing young talent like Bill Randle, Jack Paar, Soupy Sales, and Casey Kasem, all competing for a young audience. By the time a jock made it to Cleveland, if he didn't have his chops down the station management would send him packing. Northeast Ohio boasted a very sophisticated radio audience, and long-established stations like KYW (later WKYC), WHK, and others battled fiercely to keep their listeners. The 1960s, especially, were exciting for not only music but the radio stations and personalities who brought us the tunes.

When Mike's book *Radio Daze: Stories from the Front in Cleveland's FM Air Wars* (2003) was published, it sparked a lot of memories for Northeast Ohio radio fans. So many people came out to talk about radio's important role in their lives. Someone would ask at almost every signing and public event, "When can we see a book about WIXY 1260?" This book is the answer to that question.

We started this project with a simple objective: Tell the WIXY story accurately, but still show the excitement and anticipation of turning on the radio. Thanks to some very special people, we believe we have achieved that goal. Nina Wain and Mary Jo Zingale opened their homes and went out of their way to make us feel welcome, as did the late Marge Bush, who shared volumes of photographs and promotional materials with us. We caught up with Eric Stevens when he kindly allowed us to set up cameras in his home to interview him for the *Radio Daze* TV documentary, and Billy Bass braved the icy winds off of Lake Erie when we interviewed him at the Odeon for that same program.

However, very special thanks must go to three gentlemen who unselfishly shared their time and memories, despite other pressing obligations. Norman Wain, Bob Weiss, and Joe Zingale not only gave us vital information but encouraged us every step of the way, and we deeply appreciate their efforts to help us tell their remarkable story.

Mike Olszewski, Richard Berg, and Carlo Wolff

PRELUDE TO A LEGEND

Baseball great Babe Ruth said that once you got into the batter's box, you should "always go for the fence"—put everything into your swing and try for a home run every time. Ruth set long-lasting records, but a lot of people forget that he was also the strikeout king. In the mid-1960s, a trio of radio salesmen had a vision for a station that would take Northeast Ohio by storm. Like Ruth, they "went for the fence," and scored big.

The story of WIXY 1260 starts with Norman Wain, Bob Weiss, and Joe Zingale. They had come up through the ranks in broadcasting, had plenty of media savvy, and were aiming for superstardom. They attained it with WIXY and, thanks to their experience in the industry, were able to handle it. All three had an early interest in radio; their love for the medium brought them together.

Bob Weiss grew up in White Plains, New York, a suburb of New York City, where most of the network shows in radio's golden age originated. As a kid, Weiss was a fan of the radio greats, including Martin Block and *The Make-Believe Ballroom,* Art Ford and *The Milkman's Matinee,* and William B. Williams, "the best deejay ever!" He attended the University of Illinois, graduating in 1956, and came to WHK after six months in a management-training program for an insurance company in Grand Rapids, Michigan. Of his decision to leave New York after college, Weiss said, "You have to start somewhere. Radio is very much like the old vaudeville circuit. You would start small and then go on to the bigger cities. Cleveland always had a reputation for starting new trends. Everybody wanted go there. It was known as the 'sixth city' in size . . . when I started there."

Norman Wain, a native of Brooklyn, New York, majored in speech and English at Brooklyn College, where he divided his spare time between acting in college theater productions and working nights at a commercial

FM station. He started his professional radio career in 1949 at a station in Norfolk, Virginia, but worked there for only about a year before he was drafted to serve in the Korean War. During his military service, he did public relations for the Far East Network in Japan for two years, and General Mark Clark chose his *Good Neighbor Club* radio show as the best educational program among military radio shows. After his army hitch, Wain went back to New York, where he worked as a classical music announcer at WNYC for a couple of years before landing at Cleveland's WDOK as program director, where he was also a popular on-air talent nicknamed the "Big Chief" and originated the station's "good music" policy. He later joined Wyse Advertising and produced and emceed a weekly one-hour TV show sponsored by Friedman Buick. That show lasted eighty-two weeks. From there, he headed Norman Wain and Associates for eighteen months before joining the staff at WHK.

At WHK in 1962, Wain gained fame for "Teen Beat," a question-and-answer advice segment he produced for teenagers. The two-minute vignettes, which centered on everyday teen life, ran three times a day and were written by Jack Hanrahan, who would later gain fame (some might say infamy) for helping start Cleveland jokes as a writer for NBC-TV's *Laugh-In.* Attorney Wilton Sogg helped prepare the answers to teens' questions, as did John Carroll University psychology professor Nicholas DiCaprio. The segments were also heard in other cities across the country to great acclaim.

Joe Zingale's story begins on Cleveland's southeast side, at John Adams High School, where he was taught by an early inspiration he remembers only as Miss Lee. And she taught in the school's own student-run radio studio, which was set up like the pros, right down to the glass walls. In the 1950s, Cleveland's school system was one of the best in the country, but even so, it was highly unusual for a high school to have this type of resource and curriculum.

Zingale explains,

> Bill Randle put on his program that . . . he was going to create classes for interested students . . . who paid for them. It was after his regular programming, and when he found out that I was so interested and stood above the other students he told me that he was dismantling the whole class and that I was the only one he wanted to keep under his wing. He would bring in artists to the high schools, such as Johnnie Ray, Elvis, and other top performers. In that way, he taught me how

people had interest in music that spread beyond putting music on the air, and that those people learned about the artist, the history of the music, et cetera, and Bill Randle encouraged me to educate myself. . . . For several years, from the time I was sixteen or seventeen, he oversaw my development in broadcasting. He encouraged me to go to Bowling Green, because there were multiple markets nearby where I might be able to work and earn some money to help pay for my education. That, in fact, is where I met my wife of fifty years—Mary Jo—who was just a teenager in Fostoria, and so was I. But I was working as a DJ at WFOB while I was going to Bowling Green State University as a student.

From Fostoria, I went to Bowling Green on the air, and then met the manager, Jack Thayer, who went to the Twin Cities, . . . St. Paul and Minneapolis. He asked me to come along to be on the air as a deejay in St. Paul, at a station called WCOW. Although their format was country western, I refused to play that kind of music, so I went on the air playing rock 'n' roll, rhythm and blues, and jazz. The owner, Vic Tedesco, was a musician and recognized that it was something that could help his

The original "WIXY Triple Play." Joe Zingale, Norman Wain, and Bob Weiss saw tremendous success in radio sales in Cleveland and took a chance on station ownership with WFAS in White Plains, New York. (*Cleveland Press* Collection / Cleveland State University)

station, so he allowed me to do that. I got married, and we lived in the Twin Cities while I worked there, and then I got drafted into the Army. Even while away on basic training I would record my show and send it back to the Twin Cities. From the Army I sent home tapes to continue my show—recording at night while the other guys slept. Then I was sent by the Army to the Army Signal Corps in New Jersey. Once out of the Army, I returned to the Twin Cities for a short time, until I decided to move to Cleveland to be nearer to my family.

He took a job at WHK and met the two men with whom he would change the Northeast Ohio radio landscape.

They sold a lot of time and made a load of money for the station. They worked with some of the greats, too, like Pete "Mad Daddy" Myers and Johnny Holliday. Then, in 1964, with the country still suffering the lingering effects of the Kennedy assassination, a new sound burst out of England and brought excitement back into the media. In February 1964, with a historic debut on the *Ed Sullivan Show,* "Beatlemania" and the British Invasion had reached the United States. It generated hundreds of millions of dollars worth of records and merchandise, and every radio station in America was ready to be part of the excitement when the first Beatles tour was announced.

Radio wars were nothing new in the 1960s, and the fight for teenage rating numbers could often lead to cutthroat battles. A Beatles concert would be a major promotional coup, but word had come down from the Beatles' management that the group would bypass Cleveland on its 1964 tour. Norman Wain recalls the behind-the-scenes battle to bring the Fab Four to Cleveland.

MCA [the booking agency] was in charge of handling the Beatles' North American tour in 1964. . . . We were sales people at WHK, and we got word that the Beatles weren't going to play in Cleveland, and we said, "Boy! That's what we need for WHK!" So we pulled a fast one. We went down to the Cleveland convention center, and we got them to promise to rent the space to us on that open date on the tour. WKYC was there, too, and they were saying, "Come on! We're 50,000 watts!" We really didn't have a deal with the convention center. Just a preliminary conversation, and finally the guy says to us, "Look. Whoever gets the Beatles gets the space. I could care less who we rent it to."

We sent Joe Zingale to New York, and he started negotiating with Brian Epstein and the MCA people in a locked hotel room. Joe calls me and says, "Send me a telegram saying you've got the date, 'cause I'm telling them that we're the only station in town that's got that date locked up!" It was the only date they could be in Cleveland. We sent a false telegram and signed the name of the manager of Public Hall. That telegram did it! Joe walks in to the negotiations and says, "Okay, guys! We've locked up the date! You've got to deal with us. You can't deal with WKYC!"

We locked the date, and our general manager, Jack Thayer, was out of town. Jack didn't know anything about it! He came back and said, "What the hell did you do!?" We said, "Jack, these Beatles are the hottest thing going!" At the time, the Beatles were big, but they were just another act! Nobody knew they were going to be these monster stars. They had some hit records, but a lot of people had hit records.

Then I went to New York to negotiate with Metromedia's people for the $50,000 down-payment guarantee the Beatles needed. They didn't understand what we were doing. I got ahold of the guy from MCA, and the two of us walked over to Metromedia's headquarters . . . , and the guy asks, "What do we get for our guarantee?" We said, "You get the Beatles!" "So what does that mean?!" We said, "Well, nothing. They get to play for you and keep all the receipts. We make no money." They went along with it.

"The three of us had been sales people at WHK. We did real well for those times," Wain recalls.

We made a lot of money, because WHK was a hot station. For some reason, Metromedia and Jack Thayer, the manager, let Bob Weiss, Norman Wain, and Joe Zingale run their own show. We had no sales manager, and we sort of ran it ourselves. As well as we were doing, we all decided we had to own a radio station. We borrowed a lot of money, and we moved our families to Westchester [County], New York, for a whole year [to pursue station ownership there]. And within a year we found that my old radio station, WDOK, was for sale. The guy who was selling it was my original boss in Cleveland, Fred Wolf. . . . He called up one day, and said, "Nor—Mun! I have [a] deal for you! You can buy this station!" He didn't own it at the time. He had been retained by the new owners. We went up to Buffalo and made a deal, but we had no idea where we were going to get the money. We had done extremely well in the one year we had been in Westchester. That led to WIXY, because we had found a station in New York state, WFAS, that was really defunct. It had been sort of kept under a barrel by a newspaper group that was afraid of its competition, so they didn't do much with it. We took the station from $180,000 gross to $1 million in one year!

. . . On that track record . . . , we were able to go to the old Cleveland Trust and some investors, like the attorney Eddie Ginsburg, who put together an investor group that included Harry Stone from American Greetings. We eventually bought Harry out. We gave him WFAS in White Plains, which . . . was worth a lot of money because we made value out of nothing. It was a fair exchange, and he gave us his interest in WIXY so the three of us owned it . . . with our investment group, who we also eventually bought out.

Our background was sales, and we did things like managing, and programming and promotion after 5 o'clock at night, but during the day the three of us would be out there selling.

It also took a lot of confidence. Joe Zingale says, "We had a few misgivings . . . with 5000 watts and a bad signal, who wouldn't? We were not blind and deaf. We were fully aware of the challenges. A 50,000 watt clear channel was next door, paying their jocks fortunes, and we were the little guy. Ad agencies asked the same question. I answered—'Soon there will be only one rocker . . . WIXY.'" Wain also commented to the *Plain Dealer* on the purchase of WIXY and its FM sister, "When we purchased the stations, we decided to go after the masses (with WIXY) and the class (with WDOK-FM)."

Before the new station was introduced, research was vital. Bob Weiss says, "We all took our wives, jumped in our cars and went in separate directions to listen to radio in other parts of the country. The Wains went across the Midwest, Joe and his wife traveled along the East Coast, and we toured the Pacific Coast. All the time we were tuning in radio stations and writing down in yellow pads what we liked and didn't like. When we all returned, we compared notes and had a pretty good idea what we wanted our new station to sound like."

The launch date was set for the final month of the year. Even casual radio fans would find themselves in for a wild and furious ride.

1965–66

December 1965 was a month full of wonder and achievement in both popular culture and American history. Beat poets Allen Ginsberg and Michael McClure joined Bob Dylan in a "meeting of the minds" at San Francisco's City Lights bookstore. The Beatles had begun their final tour of the United Kingdom, and the Supremes, the Rolling Stones, the Righteous Brothers, and a wealth of other acts were breaking new ground with every release. *A Charlie Brown Christmas* made its TV debut. Dissent over the war in Vietnam resulted in the arrests of pacifists burning their draft cards, and the race to the moon took a giant leap as *Geminis 6* and *7* maneuvered in outer space to within just ten feet of each other. In Cleveland, a new station was on the launch pad, with the countdown under way.

Jerry Spinn, the station's first program director, said, "We studied the Cleveland market and concluded that WIXY 1260 should be designed to appeal to young adults eighteen through forty-five." The playlist included everyone from the Beatles and Beach Boys to Sinatra and Tony Bennett. And its call letters were meant to catch the ear, too: WIXY rhymed with 1260, its place on the dial, and the word "pixie," which could be used in the marketing.

Some obvious changes were in store for the longtime fans of the old WDOK-AM. WIXY 1260 premiered on Saturday, December 11, offering an eclectic mix of music it labeled "chicken rock." It was a low-key debut featuring a wide range of songs meant to appeal to everyone from teenagers to their parents, though with a serious eye toward the lower demographics. The new owners had a big job ahead of them. Wain remembers this:

> When we bought the 1260 frequency, it was WDOK-AM and FM. Frankly, the format at that time was sort of middle of nothing. We were

afraid to go rock, because we had two big rockers in the market. We had WKYC and WHK, so we came up with this format called "chicken rock," playing what we thought were the more acceptable moderate records, and we wouldn't go crazy. We didn't do a lot of promotions or anything else. We put it on the air, and nothing happened. We did some advertising. We had some bus posters and stuff like that, but nothing much happened with the station.

Wain, Weiss, and Zingale shifted programs like the *Pop Concert, Candlelight Concert,* and *Old Timer Show,* among others, to the FM band. Veteran announcer Wayne Mack was also reassigned to WDOK-FM, in a new 5 o'clock afternoon show. Thanks to the FM band, WDOK would now broadcast in stereo for eighteen hours of its day, with the remaining, overnight, hours simulcasting the mono signal from sister station WIXY. The new owners thought older fans were more likely to have FM stereo equipment in their homes, and only rarely in their cars.

A new station also brings new voices, and WIXY had plenty of them. Al Gates out of Providence, Rhode Island, was the new morning man. Rounding out the staff were Johnny Michaels; Johnny Canton; Mark Allen; and—doing a seven-day workweek's double duty on the AM and FM sides—Howie Lund, playing the *Big Beat* records on 1260 mornings and hosting the FM *Old Timer Show* on Saturday mornings. Lund was the only holdover from the old call letters, having been with the station since 1958. He was an "older guy" as far as rock 'n' roll disc jockeys were concerned, but he didn't knock the new music. He recalled that even in the 1940s rock 'n' roll was popular although it was known then as "race music." People enjoyed music with a beat, and he played to that audience. Norman Wain says of Lund's role with the new station: "We thought Howie was the only guy who had enough flexibility to be a different kind of jock. He was an older guy with a wife and a family, and I guess in those days he was maybe forty, which [for a rock 'n' roll disc jockey] was considered old. We felt we needed someone for that midday period. We just liked him a lot and felt he had the ability to get up, whereas everybody else from WDOK were 'sweet music' type jocks."

Mark Allen would later be known as Bob Dearborn on WCFL / Chicago and would go on to a long career in broadcasting. The Johnny Michaels who was part of the original WIXY lineup was not the disc jockey with the same name who would come to WIXY in 1968. The first one was really Frank McHale, who had worked at WHK for a short while in 1963 under that name. He later did another stint at WHK as Frank McHale in the late 1960s and early '70s, when WHK had a middle-of-the-road format. To further complicate matters, the second Johnny Michaels also worked at WHK, first coming to that station in 1967, when it was still Top 40, and stayed there for a while after the station went middle-of-the-road in October 1967. Al Gates was another major talent, known for his wit. ("When a boy goes to college, a father shouldn't think he has lost a son, but that he has gained back his car.")

In early 1966, Eric Stevens, barely out of high school but already a radio veteran, was one of the first staff members to sign on at the new station. Stevens started his career in 1964 as "vice president of gofers" at KYW-AM, doing any menial task that came up and eventually answering request lines for the disc jockeys. After a while, the jocks started taking him out to spin discs at local record hops, where the big-name deejay would host the dance and Stevens would play the music. Some of the jocks were in their late twenties or early thirties—way too old for high school girls—so they pushed Stevens more into the spotlight because the audience related to him. Girls would ask him to dance, and more than a few times he later received perfumed letters asking him to get better acquainted.

KYW's Jim Runyon was also doing a folk music show, called *Runyon and Folks,* and the sixteen-year-old Stevens knew enough about the music and artists to eventually produce it. Among the many highlights were his interview with a young Bob Dylan and remotes from the legendary club La Cave near East 105th Street. Runyon would credit Stevens at the end of each show, and thanks to the young producer's choice of "Here's to the State of Mississippi" by Phil Ochs, which was not complimentary of the South during the fight for civil rights, the show received at least one hate letter. KYW's powerful signal could be heard in most parts of the country, especially at night. When the courts ruled that a previous swap of stations and call letters between Cleveland and Philadelphia had to be reversed, NBC took over the newly rechristened WKYC, and Stevens was soon looking for a different position. WPVL-AM in Painesville turned him down, which turned out to be one of the biggest breaks he could get. He made an appointment to speak with WIXY program director Jerry Spinn and settled in for a momentous stay.

WIXY, wanting to get its disc jockeys out in front of the public as much as possible, jumped on the idea of the mobile studio truck right from the start. The owners signed the contract for the truck the day they closed on WDOK. Bob Weiss recalls,

WIXY on wheels. The station's satellite studio truck was a highly visible addition to its promotion. It brought WIXY's personalities to events ranging from rock concerts to store openings. (Norman Wain Collection)

We got the idea from the research trips. We figured we could take the radio station to the people. It was simply called the WIXY Satellite Studio, and it had big windows, like a car rental van, and we painted it red, white, and blue. It was fully equipped, plus it had lights on the outside that flashed with the music. It was even operational [as a studio] while [someone was] driving, and it was a moneymaker. We sold [the promotional opportunities] to advertisers to have at their place of business, and oh, yes, the jocks loved it! It gave them plenty of exposure!

There were still some problems to deal with, and one of them was the antenna signal. Even so, Wain said they were confident the station would be heard: "For its day, and remember, this was before everything moved to FM, it was not too terrible. It was comparable, believe it or not, in many ways, to WERE and WHK. . . . You didn't need that strong [of] a signal, because you didn't have the urban sprawl you do today. . . . With urban sprawl, you get neon lights and other things that emit RF signals that degenerated our signal. Burkhardt-Abrams was a big [radio-industry] consulting firm and after a while those guys used to drive in from Pittsburgh and sit on the fringe of the signal to listen to WIXY to hear what we were doing."

Needless to say, some of the older fans took issue with the change. Several listeners voiced their opinions to Bill Barrett at the *Cleveland Press:* "WIXY? 1260? PIXY? NIXY!!" "I'm fishing for a new station!" and "Doomed to failure. . . . The younger set won't listen to WIXY because it doesn't play the real hard rock, and the Vaughn Monroe set won't listen to anything that might wake them out of their trance!" Even so, their old station wouldn't be coming back anytime soon. The station also had a widening base of supporters: predictably, mostly younger listeners. One wrote, "Its music is not junk, but instead is, for the most part, high-quality popular music. The loudest records played by the station are only borderline rock and roll."

TV-radio columnist Bert Reesing wrote in the December 13 issue of the *Cleveland Plain Dealer:* "Starting early Monday morning and continuing throughout the day, dozens of perturbed listeners registered their telephoned complaints to this office concerning the swing by the station's new owners from the soft beat to rock 'n' roll."

One must also keep in mind the state of popular music at the time, with parents enjoying the music and artists they grew up with in the 1940s and '50s, along with new trends like the British Invasion. Many of those artists were enjoying great success and exposure on TV with variety shows like *Ed Sullivan* and the *Hollywood Palace,* with acts like Rolling Stones sharing the bill with crooners like Dean Martin. In truth, WIXY was playing middle-of-the-road fare from Mike Douglas, Perry Como, and Robert Goulet right along with the Supremes and the Kinks.

According to a Mediastat ratings survey for the last quarter of 1965, WIXY was entering a very competitive radio field. WKYC-AM led the pack, with WJW-AM (also with an older format) coming in second and WDBN-FM third, followed by WGAR-AM, WDOK-AM, WERE-AM, and WHK-AM. Add to these stats the fact that Medina's adult-oriented WDBN-FM was No. 1 with adults on weekends, though, and one might predict the future of radio: and FM was coming into its own.

Against all of these fierce competitors, WIXY had its work cut out for itself. Despite the new owners' enthusiasm and optimism, the station didn't get much response in its first few weeks. Wain remembers the calm beginnings, and the decision to shake things up: "We decided, 'Oh, the hell with it! Let's go for it. Let's go for the big number!' So, one day we shut down the chicken rock, and the next day we were screaming!"

Wain also recalls how in those early days, the bank's loan officer Al Brandt would meet with them weekly to see if the station was generating enough revenue to pay its bills. But more than this, he remembers how he and his fellow owners decided to take that big formatting risk.

> The three of us had a lot of borrowed money in there. Every nickel I owned was in it! The same thing with Bob Weiss and Joe Zingale. We were nervous as hell! But one morning we had the guys revved up and playing all the records we were afraid to play. The rest is history. We took off like crazy. We were lucky that we came along in a time when music was very exciting. Those records were just hot, and they helped to change the face of rock.
>
> We combined it with some disc jockeys that were so revved up that people just got excited about it.

It wasn't long before word of mouth started to generate some excitement about the daring new kid on the radio block.

In January 1966, just as the stations' new owners were finding their sea legs, an American Federation of Television and Radio Artists (AFTRA) strike hit WIXY and WDOK, and it sent personnel from both stations to the picket lines. A roster of temporary air personalities filled their jobs, which led to speculation about some of them winning permanent spots on the staff. Joe Finan was mentioned as a possible fill-in as well; the onetime TV "Atlantic Weatherman" had been stained by the payola investigations of the 1950s but remained a highly recognized and popular name in the Greater Cleveland market. A statement from WIXY management added fuel to the rumors: "We're not ruling out the possibility of his returning here." However, Joe Finan had never even appeared on the station! While the Finan name had been heard, one of the fill-ins, a member of management, said he'd used the name *John* Finan, which led to the confusion. The real Joe Finan was still at Denver's KTLN, and when he found out about the ruse, he wrote to *Cleveland Press* columnist Bill Barrett:

> I have been told that WIXY has indicated I might return to Cleveland with them. I am also told that some strikebreaker has used my name to identify himself on the air. I have no intention of leaving Denver,

no intention of returning to Cleveland. Norman Wain offered me a job which I turned down.

That Wain chose to involve my name for promotion purposes is too bad. I think this may be an indication as to management's maturity at WIXY.

I thank you for righting the record and wish NABET all the luck in the world.

In all fairness to Wain, it should be pointed out that no record exists of his suggesting Finan or endorsing the use of his name by anyone at the station. The strike left a small group of people to fill a lot of different jobs. "Joe Zingale and I had both been disc jockeys," Wain recalls, "so we both took big segments. I took a six- or seven-hour shift, and he took the same. We were there trying to keep the thing alive. It was tough, because we were trying to run the station, too, and sell time and do everything else. Bob Weiss was running around trying to keep things together. It was a tough period for us, but we eventually went into a negotiation and worked it out."

Even without the strike, WIXY's new owners faced an uphill battle. Wain puts it pretty simply: "Even if you're doing well, it takes a long time in radio to get repeat business, to get people in the habit of buying you. A lot of the agencies and their clients are in the habit of buying other stations, and we had to sort of worm our way into it. The result was even if we were doing business we weren't showing profit. It was tough. We had some very rough moments."

In a Radio Response Rating survey of the Cleveland market in February 1966, WKYC was named the top influence on pop-singles record sales. Record company executives, dealers, and distributors, and one-stop distributors and rack jobbers, who put product in stores, were all polled for this survey, which gave WKYC 39 percent of the vote and WHK 36 percent. The new kid on the block, WIXY, was third, with just 11 percent; Bill Randle's show on WERE and CKLW (the 50,000-watt powerhouse out of the Detroit-Windsor area) came in fourth and fifth. No one, Wain remembers, expected that much success in such a short time: "We were very surprised. I'll tell you, one of the things we did is we modeled it, to a certain extent, after WABC / New York, which was a monster station, and we went down to the same production house in Nashville to get our jingles. They were not too dissimilar to WABC's."

There was plenty of new music coming out, and WKYC-AM and WHK-AM were obviously WIXY's main competitors in early 1966. But how much of the new product ever made it to air at any of those stations? WKYC was running a playlist of about seventy records. The station's in-store survey was, of course, much shorter, but a radio outlet's playlist and its printed survey are not necessarily the same; a program director may decide not to air all of the records listed on the survey. Also there may be extras, "hit bounds" and LP cuts that were aired but never made it to the survey.

Each week on WHK, three new records were played Monday through Thursday, with listeners deciding their favorites. Each day's would be featured on the Friday segment, and the record that got the most votes would be declared the *Bob Friend Battle Sound* winner. The station had its Top 40 Tunedex, the weekly survey showing a song's current popularity (formerly its Top 50), with a new rundown every Friday afternoon. That could also include hit bounds, LP cuts, and instrumental extras that might also be used as lead-ins to the news, not to mention songs featured on Bob Friend's nightly *Battle Sound*.

In early 1966, WIXY aired all sixty records in its survey, and it had an ultimate song of the week and LP cuts from a featured album. Deejays would count down the entire list weekly. By the middle of 1966, though, the station stopped playing the entire list and started featuring only certain selections from songs 40–60. It just made good business sense to station management. "Believe it or not, even though we had the WIXY 60 survey, we played a very short list," Wain explains.

We had a complicated system. WABC had an even shorter list than we did. That station played twelve records. They were incredible! Our top ten occurred every hour, and the records were shorter. They were two, two and a half minutes. We were able to play the hot records over and over again. We were still playing thirty to forty records, but the top ten were coming up much more often than Nos. 20 through 25. We were very careful about how we programmed the station. The newest records were on at night when the kids were really listening, but during the day we had the established hits.

We found that we lost audience when we played the whole sixty records from the fifties, forties—we did our own research. Our audience was much more interested in No. 30 to No. 1.

WIXY debuted at the onset of the so-called high camp nostalgia boom brought on the previous January by the instant popularity of ABC-TV's twice-weekly *Batman* series. There were lots of parodies on radio, including Ted Hallaman's *Ratman and Mousie* on WGAR, and 1260's entry that April was "the most fantastic crime fighter the world has ever known . . . . *Chickenman!* (He's everywhere! He's everywhere!)" The three-minute vignettes were the product of Dick Orkin at the studios of WCFL / Chicago,

Chickenman took Northeast Ohio by storm! Here WIXY's Larry Morrow models the latest in feathered crime-fighting wear. (Photo by George Shuba)

with fellow former KYW / Cleveland jockey Jim Runyon and actress Jane Roberts, who also did traffic reports for the station. The parodies became a huge hit. Runyon was the segment's announcer and ended every episode with his trademark, "Well-l-l!" and promises of more adventures to come. In 1967, the show even spawned an LP as well as a contest to draw Chickenman. His impact, especially in Cleveland, could not be understated.

> I can't begin to tell you how important it was. Other stations would put on *Chickenman* once in the morning and once in the afternoon. Something like that. We put on *Chickenman* every hour. We played the same episode all day long, so you heard the same one twelve times a day. The next day another episode. We did what I think was one of our best promotions ever with *Chickenman,* these little yellow stickers that said, "These premises protected by Chickenman!" We put them in phone booths, toilets. Everywhere! All over town. . . . *Chickenman* really set us apart from everything else on the air. People were afraid to miss an episode! We'd have people call in and say, "What happened yesterday on *Chickenman?*"

WIXY listeners got a kick out of gimmicks like *Chickenman* but really tuned in for the cutting-edge music. In 1966, while some groundbreaking artists were producing great tunes, there is no question that the Beatles remained the top draw in the entertainment world. By spring 1966, the Fab Four had announced plans for another tour of the United States, and wheels were in motion to bring them back to Cleveland. Fortunately, Louisville, Kentucky, pulled out of its agreement to stage the group on August 14 at its state fairgrounds, for fear that fans would trample the facility just four days before the fair opened.

The Zingale-Weiss-Wain team that brought the band to Cleveland in 1964 got ready for another battle, this time with much higher stakes. Wain remembers well the challenge he and his partners faced in securing financing:

> We figured, "Hey! If we bring the Beatles in, we establish ourselves as *the* rock station in Cleveland!" Remember, WKYC and WHK had not rolled over. They were still battling us for the top position. We figured if we bring them in, we've got it. We called MCA, and they said, "The only thing that's changed is the guarantee is $75,000," which was all the money in the world! We went to Harry Stone and Cleveland Trust,

and, while he helped a lot, Harry didn't really understand it, either. He asked, "So what do we get for our $75,000? . . ." "You get promotion!" . . . We talked him into giving us $75,000 over and above any budget, and we plunked that down for the second Beatles show.

The station had a whole summer to sell tickets so they could pay back the bank, and no one seemed worried. Little did they know what lay on the horizon.

In the middle of May, station employees went on strike against WIXY again, over wages and working conditions, when members of the National Association of Broadcast Employees and Technicians (NABET) carried picket signs outside the station's business office. They got plenty of support from other unions as well, with members of Cleveland's branch of the AFL-CIO and other unions coming out in a show of solidarity. Members of all union families were urged not to listen to the station until the strike was settled, and at times it was ugly. Howie Lund was president of the Cleveland chapter of the American Federation of Television and Radio Artists, and when he refused to cross the picket line, station management fired him. He had claimed station management provoked the strike and that the dispute was "immoral and illegal." WIXY managers countered, saying they had asked Lund not to stay off the job and he had forced them to exercise their legal options to replace him—along with another holdout, newsman Julian Mouton. Norman Wain said, "We've got to be able to operate our own station. It's that simple." By the first week of June, WIXY management filed a $500,000 damage suit in Cleveland Federal Court against the Local 42 of NABET, charging the union with unfair labor practices and allegedly threatening and coercing advertisers to stop doing business with the station. That suit also requested $25,000 for each day the alleged threats continued, though a representative for the strikers denied those claims. While a spokesman said the union did ask sponsors to halt advertising, it made no threats, and all actions pursued by the union were legal under the National Labor Relations act. The strike would come to an end, but not before a move by WIXY management that drew smiles even from some of the picketers. The bosses hired three models to set up a counter picket line promoting the station's musical merits.

Lund's dismissal was a hot topic. After all, he'd been at the station since 1958, when it was still WDOK-AM. To be fair to WIXY, the station had a no-strike contract with AFTRA, which was separate from the NABET agreement. Lund and Mouton decided to honor the NABET line on

Striking engineers at WIXY and WDOK are joined on the picket line by board members of the Cleveland Federation of Labor. (*Cleveland Press* Collection / Cleveland State University)

principle, and in heavily blue-collar Northeast Ohio they received plenty of support from the public. In return for their honoring the picket lines, the reinstatement of Lund and Mouton became an issue in contract talks between the station and NABET. As it turned out, Lund was able to negotiate a financial agreement separating him from the station, and he was now at liberty to pursue other interests. His first order of business was to host a planned cruise to Hawaii. Cleveland hadn't heard the last of Howie Lund—he turned up on rival radio station WKYC late that summer filling in for Jerry G. Bishop for a couple of weeks. By the end of the year, he would have his own talk show on WERE-AM.

Deejay Jack Armstrong was part of the WIXY staff by now. "We had heard about this guy down South named John Larsh," Wain remembers. "He came up, and I said to him, 'We're not going to put you on the air as John Larsh. You're going to have to change your name. . . . How about Jack Armstrong, the All American Boy?'" "I was referring back to a radio program . . . which he'd never heard of. He had no idea what I was talking about. He said, 'Gee. That sounds okay!' To me it was tongue in cheek, and to him it was his new name!"

Armstrong dived head first into each show, cranking up his headphones to full volume to accommodate a hearing problem. And after tours of duty at radio stations in North Carolina and Georgia, he dropped his southern accent. But, Wain points out, he did not drop his high-energy delivery: "There's a record by the Rolling Stones, 'Let's Spend the Night Together.' Jackson Armstrong would scream it! 'Okay, here come the Stones, and we're going to spend the night together!' In the AM radio days, you could have the sound of the music level high, and he'd go three feet behind the microphone and scream into it to be heard, and it made it sound much more exciting because he was screaming over the record! . . . I've heard that record many times since then, and it still doesn't sound nearly as exciting as when Armstrong introduced it!"

Jack Armstrong came from a prestigious family that included a mother who graduated from college at age eighteen and a father, Dr. John Larsh, who was president of the American Society of Tropical Medicine and head of the University of North Carolina's parasitology department as well as a visiting professor at Yale and Harvard. But academia wasn't for Armstrong, who left Guilford College after a year to continue his radio career, which was really taking off. WIXY was a huge step for the former John Larsh, though he says there was still a great deal of uncertainty about coming to Cleveland.

He claims a lot of the WIXY deejays were trying to get better paying gigs elsewhere: "There was so much tape being used for . . . airchecks [demonstration recordings] that there sometimes wasn't enough left in the production room to produce radio spots!" But the upside, he said, was that the jocks were taping their best shows to land more lucrative positions. Armstrong would sometimes work seven days a week to supplement his income.

It took him a while to warm up to Cleveland. In an interview he gave to a teenage reporter for St. Michael's High School newspaper, the *Mike,* in March 1967, Armstrong gave his first impression of the city: "a dirty, backward, old town . . . with progressive people." He also noted that Greater Clevelanders embraced their radio personalities like few other cities.

The date of the Beatles show at Municipal Stadium was fast approaching, and while there were plenty of seats still available, it was likely that most fans would come out in droves at the last minute to see the biggest act in the world. Suddenly, though, it looked as if the bottom would fall out of the big event.

John Lennon had remarked to Maureen Cleave, a reporter with London's *Evening Standard,* in March that the Beatles were "more popular than Jesus now." It didn't raise many eyebrows in Europe, but when that interview was reprinted in the United States in the July issue of *Datebook,* it exploded in the media. Some stations in Cleveland and the rest of the country announced plans for a massive "Beatles boycott" and collected anything to do with the group, to burn in huge public bonfires. No one anticipated the furor that ensued, which would threaten the band's summer concert tour, not to mention the members' own safety. The worst of the negativity came from the South, while stations like WIXY, WHK, and WKYC took a more neutral stance. A Cleveland minister, Thurman H. Babbs, was anything but neutral. Babbs threatened to excommunicate any member of his congregation found listening to or even reading about the Beatles. Finally, on August 11, Lennon held a Chicago news conference to apologize in a roundabout way, saying his comments were misinterpreted. Even though he tried to make amends, his remarks hurt ticket sales in many parts of the country, and there were lots of seats still to be sold in Cleveland.

The concert was advertised with a top ticket price of $5.50, down a buck from the best seats in 1964. Stadium management didn't hesitate at all, even though the event was held in the middle of the Indians season. At first it had looked like a slam dunk that the Beatles could fill eighty thousand seats and WIXY would have no trouble recouping its down payment. However, the task would prove incredibly difficult, especially in light of the media backlash against the band. "Remember," Wain explained,

> this was the age before the mega rock shows, before the Rolling Stones were filling up stadiums. A big rock show in those days was eight to ten thousand people. We give them the money, they promise to come in, and about a month before they're here we're having trouble selling tickets! We're trying everything we can, and all of a sudden the Archdiocese of Cleveland comes out with a pronouncement that because John Lennon said the Beatles are bigger than Christ, the bishop says, "No good Catholic would go to that show!" All of a sudden, we see sales drop! We worked like crazy, and went to Erie, Pennsylvania, and Pittsburgh.

Buffalo. We went everywhere trying to sell these Beatles tickets, and eventually we sold between twenty and twenty-five thousand.

Because the media had proven so predatory, rather than stage the expected pre-concert press conference, the band's publicists decided to host an informal get-together that Sunday at the Sheraton Cleveland for invited press to quiz the Fab Four. Norman Wain was among them. Looking back, he recalls, "I was able to talk to Paul McCartney and John Lennon. . . . They were the most incredibly sensitive and beautiful people. They were absolutely stunned at what was going on. They couldn't believe the hysteria. They said, 'All we do is pluck the guitars and sing our songs, and everybody goes crazy! What did we do to get this kind of adulation?' And frankly, we didn't know either! Remember, those songs . . . weren't the beautiful things from the *Sergeant Pepper* era? It was, 'She loves you, yeah yeah yeah' and other basic stuff. They were so humble, though."

Once the tickets were sold, however, WIXY management still had to tackle the concert itself. And this was no small feat, thanks to the audience of frenzied teenage girls. Wain tells the story:

> The cops weren't that versed on how to control crowds. They didn't know what rock crowds were going to do. The cops are watching the show like everybody else, and all of a sudden, some teenage girls start rushing the stage! Bob Weiss, Joe Zingale, and myself were out there in front of the stage tackling teenage girls to keep them from approaching the Beatles. The show was actually cut short because of the unruliness of the crowd. They came back on, and the crowd got unruly again, and finally the police shut it down. We had two Cadillacs. One was going to right field, and one to left field. One had the Beatles in it, and one didn't, but the one that was empty went first and all the kids started racing after that one.

As expected, sponsorship of the show, despite the Lennon controversy and riotous conditions at the concert, helped cement WIXY's lock at the top of the ratings pile.

Controversy of a different nature arose that summer on local radio airwaves with a novelty song that made it all the way to the No. 1 spot before it was yanked over public outrage. The tune, "They're Coming to Take Me Away, Ha-Haaa," by Napoleon XIV, concerned a jilted lover whose problems with the opposite sex drove him to madness and, as he put it, "the funny farm where life is beautiful all the time." The song went on about men in white coats and weaving baskets to pass the time, and while kids enjoyed it as nothing more than a humorous little tune, those with family and friends suffering mental problems did not.

The show-business bible *Variety* blasted the tune as "the grisliest disc of the year" and said, "If this shameless exploitation of mental illness becomes a hit, it's time for us all to take to the hills." Well, it was a hit, though not everyone accepted it. Within days, WIXY pulled the song, following the lead of WKYC, which defended itself by saying even though it was in very poor taste, "We didn't play it very often." In the *Cleveland Press,* Bill Barrett countered by likening that defense to "beating your wife only on Friday nights because it's really not a nice thing to do." WHK continued to play the song until it dropped off the charts, which wasn't long after the other stations pulled the plug.

Rock concerts were still in the very formative stages in 1966, and just about anyone could promote one if they had the cash. The key word, though, was promotion, and WIXY joined with the *Cleveland Press* to host the Spirit of '66 Fun and Fashions show at Public Hall. Along with the WIXY jocks, it featured Don Webster from WEWS-TV's *Upbeat* and Ken Scott of rival radio station WHK. Today, a radio station would almost certainly demand exclusivity for an event like this—WIXY wasn't even a year old, so perhaps it hadn't the experience to consider such a tactic. The show featured a lineup that included Paul Revere and the Raiders, Tommy Roe, Chad and Jeremy, and a number of local bands, with the Saxons, Muther's Oats, the Mixed Emotions, the Outsiders, and the Tree Stumps rounding out the bill. It would also be an opportunity to see some of Cleveland's rising stars, since the Tree Stumps featured a young guitarist named Michael Stanley, who would achieve stardom on his own. And Sonny Geraci sang with the Outsiders, and Ben "Eleven Letters" Orzechowski—who would later shorten his name to "Orr" and go on to fame as a member of the Cars—was a member of the Mixed Emotions. The Spirit show also included the *Upbeat* dancers and members of the college and teen boards from Sterling Linder, Halle's, Higbee's, and other department stores.

The summer 1966 WIXY air staff included Jerry Brooke in the morning; Johnny Canton holding down the midday shift; during the afternoon drive, Johnny Walters, who left in late summer 1966, to be re-

Afternoon man Al Gates joins Jack Armstrong to plug the March of Dimes campaign. (*Cleveland Press* Collection / Cleveland State University)

placed by Larry Morrow; Al Gates in the evening; Jack Armstrong from seven to midnight; and Bobby Magic (Matchak) on the overnight beat. That September, Larry Morrow, the former "Duke Windsor," joined the station deejays. Years later, in a WIXY revival on Cleveland's WWWE-AM, he would say that he was "an all-night screamer" on the Big 8 and came to Cleveland after a staff change at CKLW-AM.

A look at the station's music survey from late summer and early fall shows a wide variety of music, with everything from Shelby Flint's "Cast Your Fate to the Wind" and the Sandpipers' "Guantanamera" to "96 Tears" by Question Mark and the Mysterians and "Shake Sherry" by Harvey Russell (an Akron Police officer) and the Rogues. Most of the stations embraced the hits, and it was up to the program director and personalities to use images and jingles to establish a unique and recognizable radio "personality" for each part of the day. To keep younger listeners, more interested in music than breaking events, WIXY introduced the "triple play," which Jack Armstrong would later claim he originated. At about five minutes before the top of the hour, the jingle would play: "Here's a WIXY triple play. Three together! O-o-okay!" Three tunes would follow, with no commercial interruption. This was actually pretty innovative: in 1966 Top 40 stations rarely played more than two records in a row. Often, listeners wouldn't hear more than one record between commercial breaks. Wain later recalled that move as "experimental, unusual" and even "radical."

"When we came up with the WIXY Triple Play," he remembered,

> we heard that . . . some stations . . . were doing a crazy thing like not talking between each record. In those days, the record ended and the deejay talked, no matter what. He told you what the record was, he told you what was coming up, and he did some commercials. That was an inviolate rule. Then we came up with this crazy idea that we were going to play three records in a row. The deejays didn't understand it. They had a lot of trouble playing three records in a row. It got to the point that Eric Stevens was told to tape the three records and put them together, and then all they had to do was push the button. . . . It took training. It was tough to do, and guys would violate it all the time. . . . They couldn't believe we didn't want something between the songs.
>
> As it turned out, that WIXY Triple Play became a very important part of our ratings. Listeners might be on another station, but on the hour they would tune it in because they could hear three records in a row, which, for its day, was a wild thing. We always did it just before the hour, which helped with the ratings the quarter before and the quarter after.

It was such a success that someone proposed the "WIXY Six-Pack"—six records in a row. The triple play was programmed around the clock, but the six-packs were only after 4 P.M., when the commercial load was a lot lighter. Armstrong remembers a lot of commercials were usually scheduled after triple plays or six-packs.

The concept apparently worked: according to the July, August, and September Hooper ratings, the station had risen to No. 1—with a 5,000-watt signal, against WKYC and WGAR's 50,000-watt signals. Although this was a stunning victory, Wain frames it realistically:

> The ratings were different then. You have to understand that there were only eight stations [of prominence] in Cleveland then. There's like thirty stations today. There was no real FM penetration, and of the eight

stations only six were playing popular music. Two were urban [actu-
ally rhythm and blues]. You really only had four competitors, and the
ratings services at that time were Hooper and Pulse. I'm not sure how
accurate the ratings were. There were some quarters where we had 50
percent share in the evening, which was amazing to me, because that's
when our worst signal was. I have a feeling that people told the ratings
folks they listened to WIXY whether they listened or not. It was the
sharpest, hottest thing in the world! How could you say you listened to
anything else?

Even with these caveats, Joe Zingale says, the management plan obviously worked.

Some things are just right. Like a good recipe, the ingredients come
together, and the final result is magic. It always seemed that way to us
and our families. We had created something our kids could listen to.
Something that we could include our families in. And that was true of
everyone who listened to WIXY. No need to be embarrassed by what
the parents were listening to.

   Although tightly controlled by the owners, there was a young, bright
and intelligent fun approach to programming and promotions that made
it easy to attract the right people. It was nothing but fun. It seemed like
it was a big family having a lot of fun all the time. The fact that management was only slightly older than the staff made it easy for everyone to
understand each other also.

Zingale also said the station benefited from very close scrutiny, giving
credit for its success to "the day-to-day, minute by minute involvement
of the owners, who really made all of the decisions regarding format,
promotions, and ideas to keep the station in the forefront." He further
explains:

Also, our broadcasting background influenced by my days with Bill
Randle as my mentor, my jobs as an R&B jock for many years before,
my years of being around promotions in several cities and the sales and
promotion experience of Weiss, Wain, and Zingale. This gave us perfect
understanding of the whole business in general. Once the formula was

honed, it was easy to follow as long as the record performers were producing good music. It was easy for the trio to keep coming up with the ideas. We knew how to sell advertising, what the agencies and advertisers needed and wanted, and how to keep the business end supporting the talent end of the business.

Although each disc jockey was in charge of his hours' record selection from the WIXY playlist, the formula every hour included five from the top 10, five from the 11–20 group, four from the 21–30 group, and one from the 31–60 group. There was also an optional album cut—or the "ultimate song"—the "pick hit" played once an hour. Oldies were also sprinkled throughout the broadcast day. Norman Wain told *Billboard* magazine, "What we're trying to do is create an 'a Go Go' image as a moving, happy operation. The station comes out of its news program with an up-tempo record and slow records are always surrounded by up-tempo records."

He also said WIXY's success was based on four important points. "We were at a time when the music was very exciting. We simply played exciting music. That's number one. Two, the jocks and the promotion. It was exciting because we had good people on the air who were into the music, and they were excited about doing things. Then, Chickenman, and fourth was the triple play. We created a situation with those four pillars where you had to listen to WIXY. If you didn't listen, you were just square!"

In the midst of WIXY's rising success, November 7 saw the debut of another key WIXY voice—"The Childe from Wilde," Dick Kemp out of Chicago. As Wain tells it, Kemp lived up to his name: "The Wilde Childe was just about the most incredible character. He lived that life. There was nothing on the air that was different than the way he really was."

Kemp had a lot of radio experience for a guy his age. Years later, he would recall his early days in radio and his arrival in Cleveland: "I started in Midland, Texas, at a little bitty station . . . a 5000 watter . . . ; it covered 300 miles in every direction. It was a daytime nondirectional and a good nighttime signal. I went to Fort Worth next, for about a year, and then I went to Dallas at KLIF for a little while. They bought a station in Chicago and . . . asked if I wanted to go, so . . . the time I started to when I went to Chicago was . . . less than two years. I was about twenty-one when I got to Chicago."

Of his nickname, Kemp explains: "Gordon McLendon, an old-time radio man [McLendon's family owned KLIF], he gave me that name. We had a lot of competition, and he was creating excitement, and we did a hell of a job there."

Kemp recalls the invitation to come to Cleveland: "When I first came to WIXY, it was a very close-knit organization. I didn't even care if I got paid or not. We just liked radio, and we made a little money, and it kept getting better as time went on. I wasn't really after the money. We started out making short money and ended up making pretty good money!"

But there was a lot of work to be done before reaping those financial rewards. That level of success took a serious and focused team effort. A cornerstone of the WIXY team was Larry Morrow. In a mid-1970s interview, "the Duker" said it was tough breaking into the business, but perseverance and a lot of luck eventually paid off:

> I attended the Detroit School of Broadcast Technique and spent about three months there. While attending . . . , I was working at General Motors . . . on the line and also . . . [taking classes at] the University of Michigan, trying to get my degree in marketing. I got my first job spinning records at WPON in Pontiac, Michigan, for a dollar an hour. Don McLeod, who used to be a giant disc jockey in Detroit, was the morning man. I went in to Don and said, "I would like a job at your radio station, . . . as a disc jockey." He looked at me and said, "You're not good enough! You don't have any experience." "I can do anything. I can do a show." He took me into a recording studio and . . . said, "Do a show!" I didn't know the first thing about it. I knew how to spin records and what a board looked like, but this board was totally unfamiliar to me. . . . But he still gave me a job spinning records.

McLeod helped Morrow get his foot in the broadcasting door, but six months later, it seemed like he was stuck with just that one foot. He remembered:

> One of the disc jockeys on remote said, "Larry. Here's your big chance. I'm going to let you read a spot. As soon as I give you the buzz, you read it." I'll never forget, it was a spot for Pontiac State Bank, and you switch the lever to the right to go on air, to the left to be on cue or audition. He buzzed me and I started talking, and when it was over the announcer came in and said, "How stupid of you! You'll never get another

chance to do anything!" I asked, "What happened? I did a good job and
. . . didn't make any mistakes." "True," he said, "but it wasn't on the air!
You threw the mic in audition."

Fortunately, Morrow didn't let this hiccup stop him from setting his
sights on bigger markets.

There was an ad in *Broadcasting Magazine* for a disc jockey at WKHM in
Michigan, and I called and told the manager I had experience. He . . .
asked me for an aircheck, but I didn't have one. I suggested I come up for
a live audition. He agreed, and it turned out to be one of those weird
auditions where they put you in a room and say, "Start talking about
yourself, for about fifteen minutes." You start saying funny little things
about your name and where you live. . . . And then I heard, "No! I want
you to ad-lib!" I did, I thought I was terrible, but he still hired me in July
1962. There was another stop after that [before] I ended up at CKLW
from 1965 to July 1966.

Morrow would also take on yet another air name.

When I worked at CKLW I never used my real name in radio. At WKHM
I was called Larry Morris or something. When I worked in Flint, Michigan, we were the "Jones Boys," and I was "John Paul Jones, your captain in the morning." . . . When I went to CKLW, . . . Bob Buss, the
general manager . . . , said, "Larry, you don't mind if we change your
name to Duke Windsor, do you?" . . . That name was like "Freddy Detroit" or "Carl Cleveland." I was a little aggravated. I started looking for
a nickname because I couldn't stand to call myself just "Duke." . . . One
day I just started calling myself "the Duker," and it stuck. . . . I brought
it to Cleveland . . . [because] when I was working between 1965 and
1966 at CKLW, WIXY had yet to be born, so by the time I got here Duke
Windsor was a popular guy.

Popular or not, Morrow learned firsthand about the volatile nature
of radio at CKLW in summer '66: "I was fired! They were just cleaning
out the radio station, and that was right after I was voted Detroit's Most
Popular Disc Jockey! Two weeks later. I was out of a job. That's when I
came to WIXY." Since Clevelanders could pick up the CKLW signal, Mor-

row was already a familiar voice in Northeast Ohio; that station even hosted events held at Cleveland's old Arena on Euclid Avenue.

Although he was already a familiar voice in the city, he experienced real culture shock when he arrived at WIXY. "It was terrible. It was unrealistic!" He later remembered.

> You come from a radio station like CKLW with over two hundred employees to a relatively new station like WIXY, and to sit there and look at the physical surroundings. They had dilapidated walls, and every once in a while you'd see a mouse running through the control room. . . . But the only thing that counted was what came out of the radio speaker. We sounded like a million dollars, but we looked like fifty bucks. But it only took a couple of years to . . . move . . . from our transmitter site on Rockside Road in Seven Hills to downtown Cleveland. Brand new studios, the most modern in the entire country. That's when I would really enjoy working there.

WIXY's state of the art facility at East 39th Street and Euclid Avenue in Cleveland was one of the premier stations in the country. (Norman Wain Collection)

It obviously wasn't the working conditions that lured Morrow from the Motor City. "I came to WIXY," he said, "because of Norm Wain." But the Duker's path to WIXY was a bumpy one:

> I was hired and fired twice before I ever got here! Jack Armstrong was fired, and they hired me to do the seven to midnight show. Then they hired him back, and I was sent a telegram not to come. A few weeks later I was given another call. . . . That time he had quit. Then they hired him back again! Finally, Norm Wain called to say, "Larry. I'm sorry for all the inconvenience, but we really would like you here. How about the ten-to-two slot?" I said, "I'd be delighted . . . but I don't know about you people. You've hired and fired me twice already and I haven't even set foot in the radio station!"

Wain remembers Morrow's eventual arrival at WIXY, and the realistic financial picture waiting for him and other deejays at the station.

> We weren't known for big pay. What we used to say is, "Look, you come in here, we're a hit station and you'll get record hops, you'll get endorsements, and you'll get sponsors that want to use your voice." Larry Morrow is a good example of what I'm talking about. We got him from Detroit, and we told him he'd do a lot of commercials. Larry had a big voice. He did half again as much of his pay in outside work. If you were entrepreneurial and looked for the business as a disc jockey you could do a lot better than your salary. Even me. When I was the "Big Chief" Norman Wain on WDOK, I was able to do voiceovers with the legendary Ernie [Ghoulardi] Anderson.

Joe Zingale would say the same thing. The management told new deejays clearly that they were joining WIXY on good faith that the best was yet to come, if they offered their best. "We often hired the least expensive young jocks we could find," he recalls, "and luckily we could do that. They included some great on-the-air personalities such as Larry Morrow, Johnny Walters, Bobby Magic, Jack Armstrong. . . . They could talk fast, play great music, and did not have to be paid too much. We wanted guys we could send out into the community on promotions and who would be presentable and personable with our listeners. We got that in spades."

Bob Weiss agrees: "We hired young talented guys who rolled up to the station with all their clothes piled in the back seat. They were young, but they were also hungry, and that's what made a difference. They wanted success."

And, given the tight budget of the young station, even the most determined deejays weren't in for nonstop smooth rides. Morrow recalls a discussion about one of his first paychecks.

> I was only making $130 a week at WIXY, down from about $300 at CKLW. When I came to WIXY I had been out of work for about eight weeks. I couldn't get stations like WMCA in New York or WCFL in Chicago to hire me full time. I had been there about a month, and Norm Wain told me they were in terrible financial condition . . . trying to finance the station and still make payroll. He said, "I don't know where we're going to get the money to pay you this week!" Payday was on Wednesday, and I think it was a Tuesday night when they got the financing to pull us through.

Despite the uncertainty, the Wain-Weiss-Zingale plan was a winning one, which included exciting the staff as well as the audience, as Morrow well remembers.

> Norm Wain was probably the single most important man in my career up until I left WIXY, because that man knew how to spread enthusiasm. He was the most positive thinker of anyone I had ever met in radio. He would say, "If you want this done the right way, do it yourself," and "If you want to be good, listen to airchecks. Critique yourself." But the most important thing that Norm Wain instilled in me was . . . that you should know the audience. The only way to really know the audience was to go out and see them. I would go to shopping centers, wherever our signal would go, and I would ask people what radio stations they listened to. If they said WIXY, then I would introduce myself as Larry Morrow. Same thing for record hops. I would do some gigs for nothing just so people would get to know me.

Morrow and his fellow jocks' efforts to know and be known by the WIXY audience certainly paid off.

1967

Popular music was a driving social and political force in America and around the world. In the first weeks of 1967, the Human Be-In, held in San Francisco's Golden Gate Park, drew international attention to that city's Haight-Ashbury scene. The brainchild of Allen Cohen, who edited the underground paper the *Oracle,* the festival spotlighted Timothy Leary, who invited festival-goers to "Turn on, Tune in, Drop out!" At that same gathering, poet Allen Ginsberg coined the phrase "Flower Power." Radio and a new psychedelic sound fueled this growing youth movement, though much of the United States continued to embrace the hits on the well-promoted AM radio band.

In this changing social climate, 1967 saw WIXY maintaining a partnership with Belkin Productions to bring some of the biggest names in entertainment to Cleveland venues. The Monkees were red hot, with a hit TV series and a string of radio-friendly hits. WIXY partnered with the *Press* to promote the show in a Monkee Business Trivia Contest, offering free tickets and records to the winners. They had the top song on the WIXY 60 list when they came to town that January, sharing the survey with songs as diverse as Prince Buster's "Ten Commandments" and the Electric Prunes' "I Had Too Much to Drink Last Night." The station's mix was certainly eclectic, recalls Wain.

> We believed in Top 40. The original concept of Top 40 is the top records from every list in *Billboard.* The black list, and the country list, whatever. We thought we should play something from every list. That's what Bill Randle was about, too.
>
> The best of all kinds of music. I'll never forget listening to the Mormon Tabernacle Choir on Bill Randle's show followed by Johnny Cash. . . . People want variety.

No appearance in Cleveland was complete without a visit to the city's rock press. Here Mickey Dolenz of the Monkees enjoys a conversation with the *Plain Dealer*'s Jane Scott. WIXY's Norman Wain can be seen in the background. (Photo by George Shuba)

I'll never forget, maybe twenty years ago. We visited China, and I was going to put together a radio program to be broadcast in China. It was going to be the *World Music Report.* We were going to take music from each list in *Billboard.* . . . I went to a recording studio and recorded it. The engineers said, "Hey! Where can we hear a program like that?"

What happened in radio is the consultants and the business people and everyone else said, "No. You've got to be known for something. You're either country, or you're this, or you're that. That's how we can sell it." That was based on a sales premise, not an audience premise. That was the Top 40 concept.

There was also a bigger potential threat on the horizon, and Eric Stevens says he saw it right away. "I went to New York and I was taken over to WOR-FM, and there was "Murray the K." I was only like eighteen or nineteen years old, and I had just been in high school a year and a half ago," he recalls. "Having been around the music business 'Murray the K' was an icon! They put the headsets on me and said, 'This is the on-air feed,' and there was 'Society's Child' by Janis Ian. I'll never forget that, because I said, 'This is the on-air feed? AM is dead. It can't compete against this!' There were people in radio who said FM would never be successful because it could never be in enough cars." They would be proven wrong.

There was also a bit of a changing of the guard at WIXY, with Al Gates and his imaginary bird, "Feathers," now doing the morning shift at 1260, but perhaps the biggest news was that Jackson Armstrong had switched over to rival WKYC. How could WIXY lose one of its franchise players less than a year after his arrival in Cleveland? Armstrong never had a contract with 1260. He had just turned twenty-one, and after a lunch with 'KYC honchos he signed on at the 50,000-watt giant. The signal strength was a factor, but Armstrong also said more money and added prestige came into his decision. He signed a two-year deal for the seven-to-eleven night-time shift. At WKYC, to avoid any legal conflict with WIXY, he used the name "Big Jack" instead of "Jack Armstrong."

In 1982, during WWWE-AM's Sixties Reunion, Armstrong talked with fellow alum Larry Morrow about his move. "That was a strange time! Norman really felt bad about it. I did, too, but they offered me more money [at WKYC], and I was working for money. I don't know about anybody else. I was trying to make a few dollars. And they offered me Jerry G. [Bishop]'s television show! They were going to make it the *Big Jack* show. Little did I know I was going to make $40 a week for that and have to work [only] five hours to do it."

Armstrong also recounted the breakneck, exhilarating pace of his gig at WKYC: "We worked six days a week and were on tape for the seventh and I did that for almost a year without a day off. It was nuts! It was still a lot of fun because 'KY had such a tremendous signal that we were the third greatest record influence in Miami, Florida! We had people all over Ohio listening to us, as well as West Virginia. I was doing hops in West Virginia."

"Monkeemania" hit Cleveland on January 15. The so-called Prefab Four drew the same crowds and frenzy as the Beatles had previously. Ten thousand people jammed Public Hall to hear the headliners, with openers Freddy Scott and the Go-Gos and Bobby Hart and the Candy Store Prophets. The Cleveland Police weren't happy with many in the crowd: girls who rushed the stage were roughly escorted from the auditorium. The published reports all clearly stated the show was sponsored by WIXY radio.

The Monkees' Davey Jones compares notes with WIXY's Dick the "Wilde Childe" Kemp during the band's first Cleveland appearance. (Photo by George Shuba)

In February 1967, the WIXY 1260 staff joined to meet the press at Burke Lakefront Airport, where their newest addition, Lou "King" Kirby, was formally introduced to Cleveland. The downtown airport location meant Kirby could make a dramatic entrance via helicopter, but the unpredictable winter weather coming off Lake Erie made it a white-knuckle landing for all the jocks on board. The former Louis Kirby Connors was introduced as "Cleveland's handsomest disc jockey," most recently a

Texan who was actually born in New Orleans and had traveled extensively throughout his youth. Kirby started out at Butler, Pennsylvania's WBUT-AM. He then accepted a job at Washington, D.C.'s popular soul station WUST-AM, but the local NAACP chapter reportedly picketed the studio when a picture of the Caucasian Kirby was printed in the local paper heralding his arrival. Not long after, Kirby found himself marooned at WNOE-AM in his native New Orleans for three days when Hurricane Betsy pounded the French Quarter.

In Louisiana, he also took part in wild radio promotions, even spray painting his body gold as a gimmick for the James Bond film *Goldfinger* and appearing that way with everyone from Bob Hope to the Beatles. He also had a crown and red velvet cape for upcoming personal appearances,

Jerry Brooke and Dick the "Wilde Childe" Kemp sift through thousands of entries that arrived daily for WIXY's Monkee Business contest. (*Cleveland Press* Collection / Cleveland State University)

thanks to former radio station KNUZ / Houston. At the end of each show, Kirby would ask, "Are you ready, my serf?" and a young woman with a sultry voice would respond, "Oh yes, my king!" Marge Bush, the station's assistant program director, later said that, surprisingly, the female voiceover artist was *really* young: "She did that when she was only fourteen years old!"

Although the press conference at Burke was a flashy way to introduce Kirby to Cleveland, as the *Plain Dealer* reported, it was also a forum for journalists to get acquainted with all the WIXY personalities. Responding to a question from a high school reporter, Kirby said he got into radio because he was dissatisfied with his job as a mechanic. Bobby Magic was studying to be an engineer when he first started, and Jerry Brooke simply answered an ad stating he would have to work one hour a day (though once he started it, he found out there were plenty more hours). Al Gates said he was a commercial artist in the navy assigned to Newfoundland when he told the higher-ups he was a disc jockey—and they believed him and put him to work on the air. Dick Kemp was a twenty-year-old working at a job building airplanes when he decided to go into radio.

One of the most interesting statements to come out of that gathering came from Jerry Brooke. When another high school scribe asked him why there were no female disc jockeys, he claimed, "Women will not listen to other women, so they're not likely to buy the sponsor's product." Program director George Brewer quickly added, "But there are more and more opportunities for women in management." For the entire history of WIXY radio, there would not be a full-time female disc jockey on its air staff.

That spring, WHK, still a player, brought in some big-name reinforcements with a jock known only as "Mighty Mitch" and, a bit later, Russ "Weird Beard" Knight, who had scored big at KLIF / Dallas. Hal "Baby" Moore would also join the lineup in what could have been seen as a last-ditch attempt to gain ground against the upstart WIXY.

When a station like WIXY scores stunning ratings victories in just over a year on the air, its listeners expect something huge to celebrate its rise to dominance, and on July 18, 1967, the station delivered. It staged its very first Appreciation Day at Geauga Lake Amusement Park, with free admission, reduced-rate ride tickets, various contests, and the WIXY jocks to greet the crowds. But the big draw was live music—the station brought a bill that included songwriters Tommy Boyce and Bobby Hart, now on their own after writing for everyone from Dean Martin to Jay

The songwriting duo of Tommy Boyce and Bobby Hart, seen here with Lou Kirby and Jerry Brooke, were longtime friends of WIXY and even filled in as guest disc jockeys. (Photo by George Shuba)

and the Americans; Every Mother's Son; Tommy Roe; the Fifth Dimension; and, in his Cleveland debut, Neil Diamond. Larry Morrow later recalled Diamond, who had written the Monkees' hit "I'm a Believer," among many others, quietly sitting by a tree at the park before ascending the stage for a dynamite performance.

Wain recalls the Appreciation Days, especially this first one, with great fondness. "That was an incredible afternoon, and we did it for a couple of years—and it got so big it soon got out of hand. I remember the traffic on Aurora Road. Candy Forest was our promotions person at the time, and she told me 'You're always disrupting things!' I told her, 'It's not a good promotion unless you stop traffic!' We had Aurora Road jammed, and we always had some big stars. Smokey Robinson and the Miracles, the Fifth Dimension, Neil Diamond . . . big, big names out there, and it was always a tremendous day."

Joe Zingale says events like Appreciation Days took an incredible amount of work and long hours to produce, but the WIXY ratings gave the station an important advantage. As he recalls, "First of all, we had Geauga Lake, and we had Euclid Beach Park to work off of for outdoor summer events. Also Public Hall for indoor events. The families were all interested, too. It was easy to get free concerts from the performers who wanted exposure on the top rock station, so they would agree to play if we would agree to promote them. We were willing, and it played on itself, benefiting all involved." But while the station was glad to host musicians at big events, it still retained high standards, notes Zingale: "Music that was accepted for play on WIXY [or at events] had to pass our

Former Brill Building songwriter Neil Diamond entertains thousands at WIXY's first Appreciation Day event. (Photo by George Shuba)

approval, making sure it was the sound and excitement needed to continue the image of WIXY being so special in the broadcasting market."

Again, management that was innovative without being overbearing paid big dividends. Years later, Dick Kemp would look back and explain: "WIXY related to the audience that they had. . . . They had good management . . . who were right on your own level. Young guys, and we could all just work together."

The management worked well with the dynamic jocks. More than this, though, the WIXY staffers worked together and supported each other. Jack Armstrong had, for instance, opened his apartment to station coworkers, Dick Kemp, engineer Jack Hooray, and newsman Mike Dix.

There was a lot of energy behind the mic at the WIXY studio—the deejays had as much fun as the listeners, both on and off the air. Armstrong says anything could happen on or off the air and wasn't surprised in the least one Sunday afternoon to walk into the studio to find Larry Morrow with a garbage can over his head "yelling about being in a well."

In many ways, WIXY was a "money machine." It had a proven track record of delivering audience to advertisers and plenty of people lined up to buy spots. Wain remembers vividly the work that went into wooing advertisers.

> Some of the things that happened at WIXY were, to me, so exciting, but . . . the average listener . . . didn't know what was going on. But anyway, we were able to play around with the programming to accommodate commercials and deals. If a car dealer needed something, we'd do it. We could do anything. . . . One day, we had a salesman by the name of Cosmo Capolino, and Cosmo comes in and says, "Norm, kiss me! I just sold the 5:30 news!" I said, "Fabulous! But wait a minute, Cos. We don't have news at 5:30." He said, "No, no, no. You don't understand. I sold the 5:30 news at the top of the rate card, non-cancellable for a year!" I said, "Oh! *That* 5:30 news!" In other words, the guy wanted the 5:30 news, and he sold it at top dollar for a year. He got 5:30 news! It was on the next day. That's the kind of crazy stuff that came out of there.

Joe Zingale likened the WIXY management to a three-legged stool combining programming, sales, and promotions. The station's very aggressive sales team included Capolino, John Rischer, Jim Embrescia, and Tom Embrescia. Bob Weiss ran the sales department, and his team would sell potential sponsors by stressing the station's large, diverse audience. Despite the huge ratings, occasionally businesses were hesitant to run ads on a station that appealed to young audiences with limited cash flow. Weiss and his team quelled those concerns by giving advertisers the jelly-bean pitch. They would bring out various-sized jars of the candy, each representing a different station in town. The beans inside were different colors and flavors, representing the diverse demographics of the Greater Cleveland audience. Needless to say, the WIXY jar was the biggest and had the most jelly beans. This pitch was so successful that, again, it was copied all over the country.

National advertisers knew WIXY as well. The Eastman Agency represented the station to national clients, and a visit from WIXY generally meant a friendly chat in a steam room schvitz, followed by a steak dinner.

WIXY's advertisers also got the bonus of a jingle written by Wayne Vaughn, who came up with some of the most memorable tunes in Cleveland radio history. He charged $1,000 per jingle—a bargain for a tune that kept bouncing around in the public's head! Local businesses had "musical signatures," and Vaughn wrote each of them after researching and compiling a profile of the business. Making the deal even sweeter, companies who committed to a year's worth of ads received their jingles for free. They were also free to use the jingles on other stations. Some, like Star Muffler, would use their signature tunes for many years.

The promotions machine was another key part of WIXY's success. The department came up with brilliant schemes to keep the WIXY name in front of the public. Promotions director Candace Forest says, "Norm usually came up with the ideas, along with George Brewer (our beloved program director and one of the best friends and colleagues I ever had). . . . Every year Norm would rent us a suite in a hotel somewhere, and we would all meet there and have these two-day brainstorms to come up with ideas. It was really creative. . . . Norm really understood creativity. He would just commit to an idea, and then it was up to us to run with it and get it done. I learned that from my work at WIXY with Norm and George. They were really brilliant at that. And they were fun to work with. We always had fun, even during the toughest projects. We laughed a lot."

Local and national advertisers could often be counted on to work with the promotions department on high-profile contests like the Community Club Awards. Bob Weiss recalls, "The station would partner with the Pick-N-Pay grocery stores, and listeners were asked to save cash register receipts. . . . The receipts would be turned over to a club or community organization, [which would] pass [them] on to WIXY, and the top three groups would get a donation."

WIXY's advertising department used some creative research methods; for example, the station hired people with disabilities who were having a difficult time finding work. "We hired folks who couldn't get around easily in a wheelchair, for example. We asked them to listen to other stations in their home, and log spots from advertisers. They listened to WJW, WKYC, and WGAR, among others, and their results would be delivered to us every Friday."

Given its incredible popularity with both listeners and advertisers, WIXY held a lot of power in getting product from record companies. Eric Stevens recalls,

> We were a radio station, not a conduit for what the music industry necessarily wanted us to play. I remember going to see the movie *To Sir, with Love* and [going] back to the station after hearing the title song. The label would only give you the side they were promoting, and I called the promotion man, Hank Zaremski, and I said, "I want the stock copy," . . . the one they sell in the store. I said, "What's on the other side of it?" Hank told me, "'To Sir, with Love.'" "Please bring it out to the station today." . . . It went to No. 1.
>
> When I saw *The Graduate,* on the original soundtrack "Mrs. Robinson" was like fifty-nine seconds long. I remember walking into the studio with Lou "King" Kirby and sa[ying] "We're going to play this over and over." He told me, "You can't play a fifty-nine second record!" "Now we are!"

The label finally relented, and the song became a huge hit for Simon and Garfunkel.

Eric Stevens explains that WIXY's sound was based on a lot more than whatever songs were nationally popular. "There were records that were big hits that didn't fit my sound, like 'Sugar Town' by Nancy Sinatra," he says. "It went in the top ten, and I just didn't play it. To me, the station was a sound, and I felt that our system of finding out what were the hits was very flawed in terms of calling record stores and getting their top twenty, et cetera. We played a sound more than playing what the industry said were the hits."

Summer 1967 brought a new voice to WIXY—Jerry Butler. He'd hit the big time at 1260 after a very successful run at Ashtabula's WREO-AM, which had a reputation as a farm team for up-and-coming air talent.

The key to success for many radio promotions was getting call letters in people's faces. Teenage numbers were very valuable; that fall WHK and WIXY both offered book covers to high schoolers. WIXY distributed its covers through Kenny King restaurants in an effort to improve business for a big-money sponsor. The covers also prominently displayed lists of WIXY promotions, space for class schedules and doodling, and a picture of Chickenman, which seemed appropriate, since Kenny King's held the Northeast Ohio distribution rights for Colonel Sanders's Kentucky Fried Chicken. WHK's covers, 100,000 of them, featured photos of the station's air staff.

In November, after the students were all in school, books covered, engineer Al Kazlaukas joined 1260. He recalls, "My duties included the transmitter watch, production studio work, remotes, and maintenance. I did a lot of maintenance when they found out that I was familiar with it. And WIXY did a *lot* of remotes. Norman Wain's policy was that the best promotion was to get out in front of the public." This didn't mean just getting the jocks out to shopping centers and clubs.

The high school audience was a highly sought prize for Cleveland-area radio stations. They didn't distribute book covers, but WKYC's morning duo, Charlie Brown and Irv Harrigan, offered to sacrifice their scalps to the schools signing the most petitions stating their loyalty to the station, dying their hair in the colors of the winning school. But even with a huge signal behind them, the jocks at 'KYC knew it would take more than a couple of dollars' worth of hair dye to beat the monster called WIXY.

1968

The year 1968 was highly tumultuous. The escalation of troops in Vietnam, civil unrest, and political assassinations all shook the United States. In the midst of the tumult, radio offered a diversion from hard realities. Diversity was also key in keeping folks tuned to a station, and the folks down the dial at 1100 knew they had to try something different if they wanted to steal listeners away from the WIXY Superman.

Sensing it was falling short in the ongoing wars, WKYC would announce a heavier emphasis on R & B in evenings. Program director Hal Moore announced a playlist of as many as fifty records every week, even though the station billed it as the Hot 100, starting out soft in the morning and getting progressively harder into "super strong" rhythm and blues by night. It also hoped to take a piece of WIXY's audience by billing itself as *Power Radio* with three-in-a-row "power plays" including an oldie called "past power." New records were labeled "potential powers" and it was all based on 'KYC packing a 50,000-watt signal, one of the strongest in the United States. The air staff at the time included Charlie and Harrigan in mornings, transplanted Milwaukeean Bob "Boomer" Branson in middays, Chuck Dunaway from 3 to 6:30 P.M., Les Sims from WPOP / Hartford holding down the fort from 10 P.M. to 2 A.M., and onetime WHK voice Pete Jerome doing overnights. Former WHK executive Dino Ianni was general manager of a lineup that promised to be a winner, and would have been just about anywhere. There was one problem and it was a big one. It had to face a superstar lineup at WIXY and battle that station on the ground with killer promotions.

Oddly enough, as closely aligned as it was with the youth of Northeast Ohio, WIXY was still pretty conservative in some key areas. Just after Valentine's Day, the station ran a Sunday-night special entitled *Documentary of a Hippy*—and clearly, the folks behind the program were not in favor of flower children. The show included a fifteen-minute interview with a New York hippie named Marcy and then a discussion between WIXY newsman Bill Clark and Lillian Greenburg of the Jewish Family Service Association. They agreed: It was one thing for the station to play some of the heavy sounds of the times, but another issue completely to endorse the lifestyle.

And the jocks at WIXY did continue to embrace the wild music, with all-nighter Jerry Butler hosting ten listeners for Arlo Guthrie's show at La Cave in February. Butler's show continued to play progressive sounds, with shows like Martin Perlich's *Perlich Project* on WCLV-FM and Ron King's *Guru's Cave* doing the same on WXEN-FM. Despite the emerging underground, the real action was on the AM band; few cars or homes tuned in to FM on a regular basis. Even so, it didn't take long for WKYC to sell out two Music Hall shows by Jimi Hendrix, even though he got little play on the AM side.

Disc jockey Mike Reineri manned WIXY's morning show for several months but would be replaced by another high-profile personality. Joe Finan, who had been on KYW-AM before the payola scandals broke in the headlines, joined the WIXY crew as the morning-drive deejay that February, though he was already a familiar voice to Cleveland listeners. He also did a lot of TV work as the Atlantic Weatherman and was said to have auditioned for the anchor position for WKBF-TV's upcoming newscast.

Steve Nemeth, better known as "Doc Nemo" on WZAK, where he bought air time for a progressive rock show, also joined WIXY early in 1968. His show, *Nemo's Nook,* popularly called "The Doc Nemo Show," moved over to WIXY on March 3 running from 11 P.M. to 2 A.M. Wain says, "He erupted! People were calling us the next morning, and it was incredible how popular he was . . . ! I was listening to the two and a half minute tunes, and I couldn't believe this was happening! Doc Nemo was the hottest thing we ever did, and then Billy Bass after that. We said to ourselves, 'The world she is a changing.'" Nemo would go on to be one of the pioneering disc jockeys at WHK-FM, later WMMS, with Billy Bass.

Former WKYC jock Jim LaBarbara also joined the WIXY ranks, replacing Jerry Butler, who went to KXOK / St. Louis.

As the station was adding to its on-air ranks, Wain, Weiss, and Zingale were dreaming of an even more powerful media empire. There was still room on the Cleveland TV dial, and a battle over Channel 19 brewed. Bolstered by the success of their radio ventures, the WIXY folks applied to the FCC for the license in March 1968. They faced a lot of competition and would ultimately abandon the TV plan, but their attempts showed how confident the station owners were in appealing to the Northeast Ohio audience.

The WIXY 60 survey became a weekly staple for many listeners, who counted down the hits and watched how their favorite tunes fared. The list came from data collected from a number of different sources, including sales in Cleveland and other nearby markets and even listener requests. Assistant program director Marge Bush and music librarian Eric Stevens compiled the information, then turned it over to the station's music committee, headed by program director George Brewer. Stevens, Brewer, and a different jock every week would listen to the new releases and look over figures on current hits, then issue the list for the week. Record retailers and the *Cleveland Press Showtime* magazine distributed it to the public. Cracking the WIXY 60, however, did not guarantee a song or its artists extensive airplay.

WIXY was known for its tight sound, which was thanks in great part to Eric Stevens. He recalls: "I think we were the first station in the country to preprogram, where the disc jockeys came in and the numbers were on the records and listed on a sheet and they had no choice in the music they played. They played the list that I gave them." Stevens continues, saying,

> I had a baby blue Cutlass convertible that I drove out to the studios when WIXY was on 1935 Rockside Road and I lived in Euclid. This was in 1968. I would be sitting in traffic listening to the ten to twelve records I programmed that I wanted to hear on the way home. The disc jockeys knew that they had to play the records in order. Of course, there were no cell phones in those days, and the way they would get me to call the studio was to play a record out of order. I knew my tempo so well that I would call and say, for example, "Childe! You played the record out of order!" He'd say, "No, I just had to have you call me," because I was kind of maniacal about the sound of the station.

Lou "King" Kirby was a budding musician and fronted a band called U.S. Male that got national attention almost from its start. The group beat out twenty-two other entries in a Hollywood national battle of the bands competition sponsored by ABC-TV's *Happening '68* show. The

The multi-talented Lou "King" Kirby felt equally at home on the concert stage and in front of a TV camera as he was in a radio studio. (*Cleveland Press* Collection / Cleveland State University)

judges were singers Tommy Roe and Jackie DeShannon and TV's former *Dennis the Menace* star Jay North. Kirby and the rest of the band went on to do a series of appearances for WIXY around Northeast Ohio while still competing for the national title and an ABC / Paramount recording contract, as well as a new car and stage equipment. Dick Clark Productions invited U.S. Male back for the finals, but they had to cover their own airfare and accommodations. As a result, Kirby and company went on a breakneck schedule of high school gigs to raise the money for the trip.

They came close, but no record contract. U.S. Male placed second in the finals for the *Happening '68* battle of the bands, but they didn't walk away empty-handed. In second place, the band won luggage and bicycles in a contest judged by Regis Philbin (then Joey Bishop's announcer), "Batgirl" (and Columbus native) Yvonne Craig, Fabian, Freddie Cannon, and Lalo Schifrin of *Mission: Impossible* fame. The band would later win a sitar, which no one knew how to play, in the Thom McAn *Sounds of India* sweepstakes.

The West Coast was seeing the rise of the new style of rock with longer, more complicated songs and dynamic performers like Jim Morrison and the Doors, Jefferson Airplane, and Big Brother and the Holding Company, led by Janis Joplin's soaring vocals. While the trend had still not reached its full potential in the Midwest, the powers that be at WIXY were paying attention. Like other Top 40 stations, WIXY had to decide how deep into the new music it was going to go, and if playing this sound would hurt its format. The music industry responded to the changing scene in different ways; for example, record companies cooperated by releasing edited versions of songs like the Doors' "Light My Fire," sans the long solos, and some stations played progressive LP cuts in the later evening or overnight.

Aiming for a young demographic, with its fads and trends, in May 1968, WIXY sponsored a campaign stop by "presidential candidate" Pat Paulsen. On the controversial, left-leaning *Smothers Brothers Comedy Hour,* the stone-faced comic announced he would enter the 1968 race for the White House and started a barnstorm tour to promote himself and the show.

Pat Paulsen's brand of deadpan politics brought him to Cleveland for a stop on his satirical presidential campaign. (*Cleveland Press* Collection / Cleveland State University)

WIXY sponsored a wide array of concert acts, including British power trio Cream. From left: Jack Bruce, Eric Clapton, and Ginger Baker. (Photo by George Shuba)

WIXY sponsored Paulsen's back-to-back visits to John Carroll and Kent State Universities on May 6. These rallies, hosted by WIXY's Joe Finan, featured Kenny Rogers and the First Edition. Throughout the summer, the station continued to help Paulsen's campaign by handing out buttons and bumper stickers. WIXY also sponsored a Music Hall visit by the British super group Cream, adding its *Fresh Cream* LP to its Super Albums category and playing the band's "Anyone for Tennis?" on the same list as Bobby Goldsboro's "Honey" and Herb Alpert's "This Guy's in Love with You."

While WIXY tried to appeal to a wide cross section of listeners, its life's blood pumped from the teenage market. The station's High School Spirit contest asked students from across the listening area to gather signatures in hopes of getting prizes and a special concert for their school. No one could have expected the results. The contest drew more than 20 million signatures, with the winner, Bedford's Lumen Cordium High

School, submitting more than 2.5 million, which was 5,227 signatures per pupil. The station sent its team of disc jockeys, now called Supermen, out to host an evening show in early May that included Cleveland's Outsiders ("Time Won't Let Me"), the Swampseeds, the American Breed ("Bend Me, Shape Me"), and Bill Medley of the Righteous Brothers ("[You're My] Soul and Inspiration"). That prize package also included a Wurlitzer jukebox to be stocked with hits by the station for the next year; principal Sister Mary Vincent pushed the buttons for the very first song. The High School Spirit contest would be a long-standing favorite. Even so, it wasn't one of Candace Forest's favorite promotions.

"Norm saw 'audience appeal' as the most important thing," she says.

> He and George both had a sort of uncanny way of knowing what was going to appeal to people who listened to rock radio. I thought a lot of the things we did were kind of silly—like that WIXY School Spirit promotion. All the kids had to do was collect signatures, and whichever school got the most got a concert thrown for them. I just thought we could have put those kids to work doing something more useful or relevant. Norm warned me that at times I had "too many principles" and that could get in the way! He was right, actually.

Not all of the station's contests had such grand prizes. For example, listeners could win two tickets to "The Mad Show" at the Hanna Theater that March by calling Dick "Wilde Childe" Kemp and giving their pronunciation and definition of "Ecch!"—the catchword of *Mad* magazine's mascot, Alfred E. Newman. The WIXY Supermen basketball team also drew huge crowds, with disc jockeys playing in clubs or in various high school gyms and packing the stands in the process.

The Greater Cleveland area could boast some of the country's best radio and some of its most discerning listeners, as well. For example, in May 1968, WIXY and WKYC were among only four stations across the country to win gold records for being the first to play Dionne Warwick's million-selling "Valley of the Dolls."

Clevelanders tuned into WIXY for news as well as music. And general manager Norman Wain believed the station had hit its mark exactly, as he told the *Cleveland Press,* "Perhaps as you . . . say, radio should show more creativity in its news coverage (although I'm not willing to concede on this point). But then the question arises, 'Creativity by whose standards?'

WIXY played a major role in breaking Dionne Warwick's hit "Theme from the Valley of the Dolls." (Photo by George Shuba)

The listening public, radio's ultimate user, seems to enjoy the way radio covers news. Listening in all age brackets is at an all-time high."

Speaking of news as part of a broadcasting whole, he also explained to the *Press,* "There's plenty of creativity in radio. Granted, all of it is not in the news; maybe more of it should be, but news is only one of the things radio does. The audience, the final arbiter, has already cast its vote in terms of the tremendous popularity all radio is enjoying today."

Management at WIXY all knew that the station's creativity and popular success came from the group efforts of its staff. Program director George Brewer wrote in *Billboard* magazine that participation on every level by every member of the staff would breed ratings: "No programmer or manager can successfully take over a market rating picture and become dominant in the numbers game without the complete support and enthusiasm of his staff. They must believe in him and his plan. They must believe that they can be a winner. They must have that spirit."

He explained, "Spirit can be generated in a number of ways. It starts

Program director George Brewer interviews Beach Boy Carl Wilson prior to a Cleveland concert appearance. (*Cleveland Press* Collection / Cleveland State University)

with respect. Your staff may respect you because of your reputation, track record, actual management experience and execution of your dedication to your task. But . . . you must be fair and honest with them. And I think in some cases, you have to reveal business information to them that many management experts say must be confidential. I believe that one of the only ways to generate real spirit is through involvement. You must involve your staff in all functions of your operation."

Brewer also stated strongly that full participation was an absolute must: "Don't let your staff get the idea that they are above certain tasks. The more your staff is involved with the actual operation and execution of ideas and projects, the more they feel part of the entire picture. The deejay that dusts off the console, takes the empties out and helps put the studio in order, helps count contest entries, along with the production, and participates in programming meetings feels that he is really part of what's happening."

Joe Finan, part of the involved staff Brewer wrote about, brought a new angle to morning-drive radio, informing and debating with his listeners as he entertained them with music. His talk-music formula worked well—so it was a cause for concern when, in June 1968, ads in Denver newspapers boasted, "Joe Finan Will Return." Finan had just come back to Cleveland from a visit to Colorado and went on record saying he had no intention of leaving Ohio "for now at least."

Summer was huge for WIXY and other Top 40 stations. Kids had a lot more time to listen to the radio, and the station had a lot more opportunities to sponsor outdoor events. Chippewa Lake, a heritage amusement park in Medina, featured live animals as part of its attractions, but its biggest draw was American Indian chief Wachickanoka. In a promotion with 1260, the chief vowed to stay in a "WIXY snake pit" twenty-four hours a day for a month to break the old record, twenty-one days. He would also handle poisonous snakes in scheduled demonstrations for the public, and the pit was open for viewing every hour of the day. WIXY also offered a $5,000 prize to anyone who caught him outside the pit during that time.

And the station hosted many other wild events. Lou "King" Kirby sent out thousands of cards for his fan club and rode in a psychedelic bathtub on Lake Erie and sent back boating reports. In July, Joe Finan flew high over Parmatown Mall in a hot air balloon, accompanied by music from the Selective Service band and a calliope. It was an all-out radio war on land, in water, and in air, and the station never let up on its competitors.

Finan's hot-air-balloon antics were a great visual promotion; Wain still considers them among his most memorable moments. "Favorite?! I'm glad we lived through it! As Joe Finan was going up in the balloon, our lawyer turned to me and said, 'Norm. Do you have insurance on this?' I said, 'Insurance? What's that?' He said, 'What if he hits an electrical wire?' Sure enough, as he's leaving the shopping center, there's a great big electrical wire, and thank God he got above it."

The stunt didn't go off without a hitch, however, as one Willoughby Hills attorney learned firsthand, when he saw long-haired kids swarming his property, chasing a balloon that made an emergency landing in his backyard. He was not amused. He later told reporters that he spotted "bearded guys with black leather jackets." He exclaimed, "They were all over the place—running their motorcycles across the lawn, driving their cars off the driveway into the flower beds. I grabbed a gun and came charging outside. So would you. I thought it was a riot." Aboard that balloon were Joe Finan, two pilots, and the *Plain Dealer*'s Karl Burkhardt, who were forced down by inclement weather and low fuel.

The attorney was seriously not amused: "This was invasion of property. It aggravated me. I can understand the crash landing—I'm an aviator myself. But the gang in my yard was something else. They overran the paddock out back and scared the horses. These are yearling colts in training getting ready for the track. They shouldn't be scared. The colts were screaming and rearing and banging around." He also complained about theft and vandalism: "Someone walked off with a sander I had there in the driveway," he said, "and someone else apparently grabbed the dog belonging to my caretaker. The poor lady has been crying for two days."

The situation became even messier when rain caused a muddy traffic tie-up near the house. Willoughby Hills police officers had their hands full trying to clear it out. As many as two thousand onlookers saw the balloon's launch when it left Parmatown, and many just decided to follow it across town. Candace Forest, who helped put that event together, recalls, "I was afraid to go into the office, because I saw the story about the lawsuit on the front page. When I got to work, Norm was beside himself with joy and patting me on the back. I guess that was when I learned that any front-page story is a good story!" The attorney eventually decided not to sue, and the staff did recover after the hype.

There was tremendous pressure to keep the station's momentum going. Wain remembers the frenzy and how he, Weiss, and Zingale saw in it the beginning of the end.

> Every month we wanted to do better than the month before. We wanted to keep costs down and keep business coming in. Ratings had to go up. There was a lot of pressure, and that's one of the reasons we sold out relatively quickly—because we could see we weren't going to be able to sustain it for any really long period of time.
>
> Also, one of the biggest pressures . . . and we were early to recognize it . . . was . . . that FM was coming up. We ourselves were shocked. We had the license to WDOK-FM, and we didn't have time to fool around with that so we had Wayne Mack put together these long reel-to-reel tapes on an automatic machine that was in a closet. This machine . . . was starting to get ratings that were rivaling WIXY! It was coming up like gangbusters! We called it "Beautiful music for the lands of the Western Reserve." . . . One day, Bob, Joe and I sat down and said, "Hey guys. If the automated machine in the closet that we talked about for twenty min-

WIXY's 1968 Appreciation Day show at Geauga Lake offered a wide range of acts through the day. Lou "King" Kirby spends a few quiet moments backstage with members of Jay and the Techniques. (Photo by George Shuba)

utes a week is doing as well as we're doing at WIXY, and we're devoting day and night to that, something's happening here."

Bob Weiss remembers WDOK's "short vignettes . . . like snapshots of places like Amish country, twenty or thirty seconds long" that "gave you a warm feeling about the Western Reserve." Veteran broadcaster Tom Armstrong from WGAR would eventually be added to the staff of WDOK, settling in for a long run in morning drive.

One of the keys to WIXY's broad appeal was how accessible its jocks were. The station promoted itself with its personnel, who could be seen at many venues and events. For example, Larry Morrow hosted teen dance nights for ten weeks at Cain Park in Cleveland Heights, and other jocks often hosted two or three high-profile events in a single night. But those were small potatoes when compared to the major gig of the summer—the all-day Appreciation Day show at Geauga Lake Amusement Park. The 1968 event drew 120,000 people, to hear a lineup that included the Amboy Dukes, the 1910 Fruitgum Company, Jay and the Techniques, the

New Colony Six, Peppermint Trolley, the Box Tops, and Gene Pitney. Eric Stevens, the station's music director, organized the show, which started at 10 A.M. with remote broadcasts. This was one of the biggest crowds to ever turn out for a concert prior to 1969's Woodstock, held in Bethel, New York. The post-show traffic jam confounded police for hours and didn't clear up until 3 A.M., well after the concert had ended.

And an event of this size was not without its logistical and mechanical glitches. Candace Forest recalls, "On the day of the event, I was told to ride on the bus with Gene Pitney to the show. . . . The bus broke down on the way. It was really crazy. I don't even remember how it all got sorted out, but somehow or other we got Gene to the park. . . . I remember him as being extremely quiet, and, of course, he couldn't have been very happy about getting stranded out in the middle of nowhere in the bus."

WIXY provided crowd members with special chartered buses as well as free parking and admission. The station also used the event to work for a greater good; it set up volunteer booths for its American Lebanese Syrian Associated Charities (ALSAC) drive to benefit St. Jude's Hospital for Children. ALSAC signed up tens of thousands of people to march and collect money for leukemia research. Each volunteer received a special WIXY prize for signing.

WKYC was fighting a fierce battle against WIXY. With little chance of matching the success and crowds of WIXY's Appreciation Day, it sponsored a Teen Fair with the *Cleveland Press* at Public Hall, August 9–18. It featured some of the area's best bands, including Cyrus Erie and the Damnation of Adam Blessing; live broadcasts; and plenty of handouts and contests. The station publicized the event with a TV special featuring Marvin Gaye and Youngstown's Human Beinz. To counter WKYC's nine days of publicity from the fair, on August 14, WIXY's Joe Finan walked all the way from Akron to Cleveland's Public Square, vowing he would stay on the road as long as phones kept ringing with pledges for St. Jude's Hospital. Larry Morrow joined him along the way, and a party waited for them at journey's end, hosted by *Cleveland Press* editor Tom Boardman, who was also chairman of the local ALSAC campaign.

WIXY was the undisputed king of the airwaves in September 1968, though one of its jocks had left to join the upstart experiment with progressive rock on the FM dial. Steve "Doc Nemo" Nemeth was hired on at WHK-FM, soon to be WMMS, to work the late-night, 10 P.M. to 2 A.M., shift. It brought an end to an often tense relationship between Nemo and his producer, Barry "Buttons" Weingart, and station management. Personnel director George Brewer later said that in the show *Red, White and Blues,* Jim LaBarbara, who replaced Nemo, went on to play progressive rock standards, like Jimi Hendrix's "Purple Haze," rather than more obscure tunes. And the audience preferred the more familiar music. But Brewer also advocated incorporating some of the new progressive sounds into the regular AM format, especially if there wasn't an FM outlet devoted exclusively to underground rock. The station also lived up to its commitment to dabble in progressive music during the day by playing songs like Big Brother and the Holding Company's "Piece of My Heart" and an abbreviated version of Iron Butterfly's "In a Gadda-Da-Vida."

Meanwhile, Lou "King" Kirby was hoping to broaden his own career in television and live music by auditioning bands to perform on his proposed teen show on WUAB, Channel 43. He was also inviting attractive young college girls to apply for spots on a "stump the panel" segment of the show—by sending in photos of themselves. The *King Kirby* TV show debuted in late August, and, like its rival *Upbeat* on WEWS, offered musical acts like Dionne Warwick, B. J. Thomas, and the Avant Garde, as well as features such as candid films of Beatle Paul McCartney. *Upbeat,* more firmly established, was syndicated, but Kirby had the power of Northeast Ohio's most popular radio station and what amounted to a four-hour commercial during Kirby's nightly show. The shows went head to head at 6 P.M. every Saturday, and their battle was closely watched by the audience and management at both stations.

WIXY also incorporated a new public service plan in August when it offered a "dropouts anonymous" program aimed at teenagers with emotional or psychological problems. On weeknights, counselors affiliated with groups like the Ohio Guidance and Personnel Association and the Manpower Commission stood by phones from 7 to 10. They answered anonymous questions regarding just about any topic and referred callers to the proper agencies.

In 1968, racial tensions were very high across the entire United States. The assassination of Dr. Martin Luther King Jr. and the lingering specter of racial discrimination fueled riots in many urban areas. Cleveland experienced this firsthand July 23–28 in the city's Glenville neighborhood. Seven people, including three police officers, lost their lives to gunfire, and Lou Kirby promised to donate $200 to a fund for the widows and children of police officers killed during the riots if he failed to hit the

No one anticipated the response when thousands unexpectedly shut down East 9th Street and Euclid Avenue for a glimpse at contestants in WIXY's Francine pageant. (*Cleveland Press* Collection / Cleveland State University)

fence during a game between the WIXY Supermen and the Parma Jaycees. It may have seemed like a promotional effort built on tragedy, but the people affected most by the violence welcomed the money.

WIXY also offered distractions from the serious civil rights unrest felt throughout the country. Longtime Cleveland radio fans still talk about the September 25, 1968, promotion, particularly. Staging this event was a gutsy move, considering the call for women's rights and equality had been gathering momentum throughout the country, especially that year. WIXY's stunt was inspired by a woman named Francine Gottfried, who literally stopped traffic on Wall Street when she walked to lunch from her job at New York's Chemical Bank.

Word of her "attributes" quickly got around, and it wasn't long before the amply endowed Gottfried started to gather crowds. Newspaper reports say as many as five thousand people a day gathered to see her step off the BMT subway ramp, matching the crowd for the 175th anniversary of the New York Stock Exchange. (That event featured Vice President Hubert Humphrey, Governor Nelson Rockefeller, and New York mayor John Lindsay.) She even made it to the pages of *Time* magazine. The WIXY promotions department sensed an opportunity and put wheels in motion to find Cleveland's answer to Francine. The station invited girls to compete for the local title and instructed them to show up at the corner of East 9th Street and Euclid Avenue, one of the city's busiest intersections, during lunch hour. No one anticipated what would happen next. It all started with a morning newspaper. Joe Zingale explains that after his wife, Mary Jo, read a *New York Times* story about Gottfried and told him about it, "it was a natural to have a contest for Cleveland's pretty buxom girls to walk down Euclid Avenue and 9th Street to see what would happen."

Norman Wain remembers that day well. "I'll show you how unstructured and unbusinesslike the Francine Gottfried promotion was," he says. After Zingale called him about the *Times* story, the whole team went to work very quickly. "I wrote a commercial, ran down to the station, and got the guys to record it. We were on the air like twice an hour with this thing that Friday morning, all day Saturday, all day Sunday. The ad said, 'Meet us at noon on Monday, and we're going to find WIXY's answer, Cleveland's answer, to Francine.' I have to tell you, we didn't know if it was going to work or not, and it was so stupid that we did it so quickly over the weekend that no one even thought of what the consequences would be."

Monday morning, Wain, Weiss, and Zingale drove downtown, only to have the police stop them at 18th Street. Wain recalls the morning: "We parked the car and ran down to the corner of East 9th and Euclid, and as we came down we saw huge crowds all over the place. Guys were leaning out of the window of the Union Commerce building, the Scofield Building, all those buildings down there. . . . We didn't have a stage. We didn't have anything set up. We found an AT&T truck there, and we told the jocks to jump on top of the truck and we'll do it from there." The weight of the jocks partially caved in the truck's roof.

Wain and the rest of the WIXY crew had no idea what a huge event they'd launched. He recalls: "We had been nervous whether any girls would show up, so we arranged for a couple of models to be there. Sure enough, there's some girls and two or three jocks, and Jim LaBarbara has a measuring tape. The crowd is just incredible! Traffic is stopped in four directions! People are jamming the streets! A cop comes by on a horse and says, 'Hey! You got a permit for this?!' I told him, 'Officer, how do you get a permit to throw a riot!?' It was the wildest promotion we ever did. I called the papers that morning to tell them about it. I have a theory about spontaneous demonstrations. They have to be well planned!"

Zingale later remembered the Gottfried promotion; afterward, "the whole country was aware of WIXY and started to tune in to see what we were up to and what promotion we would come up with next." An estimated ten thousand people (mostly men) showed up for the event, with some carrying signs with slogans like "No Place for Twiggys Here!" The station's judge, Jim LaBarbara, chose two winners: twenty-five-year-old Sheila Moore and twenty-one-year-old Suzanne Zolkowski, each of whom matched Francine's forty-three-inch bustline. Sheila said she signed on to compete because, "I have a lot of talent in that area." Both women thought Francine Gottfried was overrated. Suzanne commented, "She's no big thing. I thought she was a little fat!"

"We weren't going to have two winners," Wain explained, "but our judge's hands were shaking so much he got his measurements garbled. All he knows is that both girls are about forty-three." Each winner was awarded a sweater and a trip to New York's Wall Street to see if she could evoke the same level of reaction as Gottfried had.

This type of promotion obviously upset the wives and girlfriends of the WIXY jocks. Jim LaBarbara was thought to be single, so the station brass tapped him to measure the contestants' "assets." When interviewed by a *Plain Dealer* reporter about his responsibilities, LaBarbara was almost speechless, his only comment, "Ub, ub, ub, ub, ub, ub!" After the smoke had cleared, it was evident that WIXY had scored yet another major promotions coup. Zingale believed that a magic formula made things happen. "With only three young owners, with a sense of what was actually going on in the world, and a young staff with young and vibrant ideas, it was easy to make instantaneous decisions and put them into motion on the air immediately. No matter how bizarre the idea sounded," he said, "if it was current and the thing seemed right, we did it. Whatever was current and happening was noticed and exploited to bring the world and WIXY into the same orbit."

Not everyone at the station favored the Gottfried spectacle. Promotions director Candace Forest now says, "I organized that event, and it really turned my stomach. It wasn't long after that that I left the station. That promotion was Norm's idea, and it wasn't one I was in favor of. I thought it was in outrageous bad taste, and as someone on the forefront of the fight for women's liberation and equality, it just irritated me to no end. But it stopped traffic and probably got us more listeners, so who's to say. Anyway, Norm wanted us to never get too far away from things that would appeal to the general public's kind of basic instincts. I guess he was pretty right about that, judging by the ratings we got."

While WIXY and other radio stations were hatching wild promotions spectacles, much of the United States' attention was taken up by something much more serious, indeed, deadly. The Vietnam War was daily being played out on TV and radio and calling young men to service. Most of those drafted, along with many of their loved ones, were in WIXY and other popular stations' prime demographic. That October, in what might have been called a public service, 1260 offered female listeners fifteen-minute phone calls to their boyfriends in 'Nam if their names were on one of four postcards drawn that month. One name was picked every week, and, as could be expected, postcards flooded the station.

The WIXY promotions department had its hands full with projects silly and serious, and at the same time, the station was busy with straight music projects. WIXY was also planning its third *Super Oldies* album, to be rushed out in November, to take advantage of the Christmas shopping frenzy. This time around, proceeds would be directed to Camp Cheerful, a local camp for children with disabilities.

Morning man Joe Finan found himself in a difficult situation that same month when a listener called in to report her son had been assaulted by "a gang of young negroes." This call took place, of course, during the very tense aftermath of Dr. Martin Luther King Jr.'s assassination and the continuing struggle for civil rights in a very racially polarized America. Finan would tell the *Cleveland Press,* "She suggested that part of the West Side was under attack" and that "Shuler Junior High and John Marshall High School were practically under siege." Those were strong words, and Finan didn't edit the names of the schools before they made it to air. WIXY held tremendous power as the city's ratings leader; thus, one upset listener's call spurred crowds of people to take their kids out of classes at those schools and left the surrounding neighborhoods on alert for any possible trouble. Station newsman Bill Clark followed up on those reports, finding out from Cleveland Police that there had been some fights at Marshall, though the principal stressed they did not result from any racial unrest. Even so, comments like those from the upset caller resulted in a wildfire of speculation and rumor, though even the principal commended WIXY on the way the story was handled.

Finan defended himself for having allowed the woman's comments: "These people call constantly, articulating their fears. I try to present the facts based on our investigation. I don't feel responsible for what happened out there at the schools. That whole neighborhood was upset all last weekend with rumors flying long before the [October 21] program." As a member of the press, he said, "I feel I have a responsibility to debate these rumors. I feel I acted in a responsible fashion."

Joe Finan had received some criticism for not editing the comments; some people even accused him of potentially setting off a highly emotional outburst. Others commended his frank presentation of the issues. The controversy quickly faded from local airwaves but not before all sides learned a lesson about how far rumors could go in affecting the fabric of society when presented on a widely heard forum.

WIXY's promotional machine was a juggernaut, bringing listeners a new offer or contest nearly every week. Prizes awarded could range from gifts as small as free LPs and food certificates—for listeners whose cars were spotted with a Manners Restaurant "'Tenna Topper" (an orange Styrofoam ball placed on a car's aerial)—to a drawing whose grand prize was a $20,000 house in North Olmsted—which drove listeners to 1260 on the dial and the sponsors for the ultimate payoff. Bob Weiss says the sales and promotions mix always generated not only plenty of excitement but also revenue for the sponsors and the station. Other radio markets later copied some of these promotions, like the Manners partnership, with equal success.

One of the popular promotions of late 1968 was 7-Up's Uncola Underground, which, in exchange for a postcard, offered a spy kit and membership card—with a grand prize of a unicycle. Anything WIXY promoted always got a huge response. Near the end of 1968, a contest was held awarding 1969 records to the first person catching a WIXY jock playing fewer than fifteen records in an hour. Granted, songs were shorter in those days of AM radio, but listeners still began to speculate that WIXY was speeding up its songs to fit in all the music it claimed. Cleveland's other radio stations also threw grand promotions. WKYC cosponsored the Miss Teenage Cleveland contest with Higbee's department stores, among other high-profile events. If anything, competition made for greater innovation in new promotional campaigns.

By November, a controversial artist, the Legendary Stardust Cowboy, had emerged, with a cacophony of sound entitled "Paralyzed." This song, if it could be called that, was nearly unintelligible screaming along with a loud guitar. Not one to shy away from controversy, Joe Finan featured it prominently on his morning drive show, taking calls from outraged listeners who called it either sonic garbage or a work of genius. Not long before, he had championed Jeannie C. Riley's "Harper Valley PTA," about a divorced woman pointing out the hypocrisy of small-town critics. However, "Harper Valley" was an actual country song. "Paralyzed" might best be described as an audio train wreck. The Legendary Stardust Cowboy would get plenty of TV time as well, nationally on shows like *Laugh-In,* and on local programs such as *Upbeat* and the *King Kirby Show,* not to mention a spot in the upcoming WIXY–May Company Christmas Parade. This first annual parade was a jewel in the WIXY promotional crown. Joe Zingale remembered the parades clearly: "The parades were done WIXY style, with easy cooperation from people wanting to be part of anything that WIXY was doing. WIXY was able to bring in up-to-the-minute personalities who drew crowds, and, since the whole family listened to WIXY, it was easy to persuade the parents to bring the kids downtown."

And boy, did it ever bring people into the city! This high-profile event drew more than ten thousand people to downtown Cleveland that year,

Lou "King" Kirby entertains fans at the annual WIXY–May Company Christmas Parade. (Photo by George Shuba)

the first time it was staged, and was even televised on WJW-TV. Channel 8's Jenny Crimm, before her tour of duty at WEWS, and *Adventure Road*'s Jim Doney hosted that first parade.

The 1968 parade featured Chagrin Falls' favorite son, Tim Conway, as the grand marshal. Twenty-five balloons along with many floats, drill teams, and marching bands added to the excitement. The celebrity lineup included singing stars Bobby Vee, the Lettermen, O. C. Smith, the Ohio Express, and, for the little kids who braved the cold, the Chipmunks. It also featured every WIXY jock, with Lou "King" Kirby and his court in a psychedelic Cadillac, the "Wilde Childe" Dick Kemp in an Amphicat, and Johnny Michaels zipping along in a dune buggy. More than three thousand people marched in the parade that first Saturday after Thanksgiving, with Olympic stars Madeline Manning, Jenny Fish, and Eleanor Montgomery representing the Cleveland Recreation Department. The WIXY–May Company Christmas Parade would be a massive event to stage, but worth every bit of effort for its promotional value alone. That year, the first place float was the Liberty Records entry, featuring Bobby Vee and the Chipmunks. Second place was a tie between Steve Popovich and the Columbia Records float, boasting O. C. Smith, and the Capitol Records float, carrying the Lettermen. Events like this, as Bob Weiss notes, could show a huge audience to current and potential clients.

But not everything WIXY touched was gold. Kirby's Saturday show was cut to a half an hour in December in its continuing battle with WEWS's *Upbeat*. With syndication, the Channel 5 show still had plenty of clout and was the first stop for guests traveling through Cleveland. Plus, a WIXY-sponsored concert with the Crazy World of Arthur Brown was also cancelled, because of low ticket sales, most likely because Brown was a one-hit wonder with his tune "Fire."

The sound of the station was a critical factor. Engineer Al Kazlaukas explains,

> WIXY used the CBS Audiomax / Volumemax combination. They were the early models, not the later 4000 or 4400 series units. We later bought a stereo set for WDOK. The reverb unit was an old Hammond [organ] unit. Sometimes we would whack the unit with our fist, and it would make all sorts of strange sounds on the air. Finally, the chief engineer, Ralph Quay, figured out what was going on and raised hell about it. Needless to say, the threatening memos came out. Ralph had

come from WGAR in the heydays of radio, when they were a CBS network hub . . . at the top of the Statler-Hilton Hotel. He knew what live radio was about.

Kazlaukas also fondly recalls a holiday remote from a very unlikely place. "We did a remote on Christmas morning from Joe Finan's living room in . . . '68. It was a hoot. Plenty of food, booze, et cetera." The jocks were having as much fun as the audience!

1969

WIXY's year-end show—which ushered in the new year by running from December 31, 1968, until January 1, 1969, playing the year's top hits—illustrates the wide variety of popular music that shared time on the station. It was a musical smorgasbord that ran the gamut—from acts like Sergio Mendes, the 1910 Fruitgum Company, and Bobby Goldsboro to the Troggs, Iron Butterfly, and the Beatles, that year's top hit-makers with the double-sided smash "Hey Jude" and "Revolution." The Beatles may have been topping the charts as a group, but a solo effort by John Lennon was getting lots of negative attention that week. Newark, N.J., police confiscated thirty thousand copies of Lennon's collaboration with Yoko Ono, titled *Unfinished Music No. 1: Two Virgins*. A nude photo of the two on the cover was thought to have violated New Jersey's pornography laws.

Chuck Dunaway debuted on WIXY on February 3, 1969, with a high-powered show featuring everything from Bob Dylan to Bobby Hebb. The show debuted with a sound of its own; the rest of the WIXY jocks used a heavy echo generated from the Fairchild Reverbatron Plate, which, part of the station's image, was strangely absent from Dunaway's first show. While its sound was different, Dunaway's show had all the other signature elements of a WIXY program. For instance, he played six-packs, "super oldies," and a generous number of spots for products, like Kentucky Fried Chicken offering five pieces for a dollar.

Also that February, the 50,000-watt WKYC-AM put up a white flag and changed to a middle-of-the-road format, playing music appealing to an older demographic. Deejays Specs Howard and Bob Cole stayed with the staff, and Fred Winston announced he would be heading to Cheyenne, Wyoming, for a vacation. The only competition for young music left on the local AM dial was WJMO-AM and WABQ-AM. Willoughby's WELW-AM had switched to a Top 40 format as well, but its signal didn't cover all of Cuyahoga County. The giant CKLW was still booming in from the north, but even that flamethrower would be hard-pressed to face off against the WIXY machine.

While new music was the keystone of the WIXY air sound, the station was also doing very well with its older catalog. Its *Super Oldies* album compilations flew off the shelves at local record stores, and weekends devoted to past classics drew heavy interest from the listeners. In fact, Candace Forest, in promotions, said the LPs were selling in the tens of thousands.

With the demise of WKYC, little challenge from the FM rockers, and ratings as strong as ever, WIXY decided to stage its annual Appreciation Day show on January 31 at Cleveland's Public Auditorium. This year, tickets were $2, but show-goers could also win seats in a special VIP section. Station ads for the show boasted that the Friday-night concert offered, "A million dollars worth of entertainment," and the lineup lived up to its billing. It included headliner Neil Diamond, Tommy James and the Shondells, Canned Heat, the First Edition, Jay and the Americans, and the Bob Seger System. Tickets were available at the venue or at Burrows book stores. Puffing out its chest a bit, the station printed on every ducat, "Thank you for making WIXY the ONLY popular music radio station in Cleveland!"

Also in January, the powers that be at 1260 got government approval to buy WMCK-AM, 1360 on the dial, in McKeesport, Pennsylvania, a Pittsburgh suburb. The 5,000-watt station could be heard in various parts of Pittsburgh. It was pretty easy to see what direction the station was taking as well; the Westchester Corporation, the name Wain, Weiss, and Zigale formally adopted before they bought WDOK, applied for the call letters WIXZ.

WKYC began its format transition by adding some new jocks: Eric St. John from WIBC / Indianapolis, and Ted Lux, who would remain a familiar and popular voice on Northeast Ohio radio and TV into the twenty-first century, from WJAS / Pittsburgh. Even though the two stations were no longer playing the same format, the folks at WKYC had a tough challenge ahead of them—a WIXY lineup that included Joe Finan in the mornings, Larry Morrow middays, Lou Kirby until 5 P.M., when Dick "Wilde Childe" Kemp held court, until 9 P.M. Chuck Dunaway took it until 1 A.M., and Jim LaBarbara held the listening audience until morning. These were heady times for the 1260 staff, with Kirby tooling around town in a 1965 Cadillac limousine and LaBarbara driving a brand-new 1969 yellow Corvette Stingray convertible.

A number of motorists on Cleveland's West Shoreway reportedly registered complaints about a WIXY billboard showing the unclothed back of an attractive young woman with graffiti scrawled across her body. The station announced that, for the sake of drive-time safety, it would paint a bathing suit or even a fur coat on the model along with the one on another billboard, near Cleveland Hopkins Airport. However, the same sign at East 96th and Carnegie remained untouched, supposedly because there were no complaints about that particular advertisement.

Public outcry forced WIXY's owners to paint a bathing suit on the back of a nude model on the station's billboard. (*Cleveland Press* Collection / Cleveland State University)

Another change, affecting one of the station's major voices, was in the wind that February. To bolster its ratings, Dick Kemp was reassigned to the new WIXZ / 1360—a gutsy move by management, as the "Wilde Childe" was a very popular personality in Cleveland. His last day on Cleveland airwaves was scheduled for February 21, and to keep his fans listening WIXY scheduled songwriters Tommy Boyce and Bobby Hart to take over the first evening shift after Kemp's departure. The station hoped to have a different celebrity fill-in every night until a replacement was announced. Boyce and Hart were in town for an appearance on WEWS-TV's *Upbeat* show and were part of a campaign to lower the voting age from twenty-one to nineteen. It was called "L.U.V."—for "Let Us Vote"—also the title of the Boyce and Hart song that was its anthem. Just a week later, singer Tiny Tim cracked the mic at WIXY as a guest jock.

WIXY needed a powerhouse voice to replace Kemp, so it quickly added fast-talking Chuck Knapp, "Your Buzzin' Cousin" from WRKO / Boston.

Laugh-In favorite Tiny Tim rode the popularity of his song "Tiptoe through the Tulips" to a fill-in spot on WIXY's evening lineup. (*Cleveland Press* Collection / Cleveland State University)

"Your Buzzin' Cousin" Chuck Knapp had big shoes to fill when he took over the spot vacated by the popular Dick Kemp. (*Cleveland Press* Collection / Cleveland State University)

Another radio junkie, Knapp would say he was destined for a career on the air. "I was eight years old when I started listening intently to what was going on between the songs. I was fascinated! Plus, I grew up in west-central Minnesota listening to Indians baseball games on the radio from a faraway station in Ohio. I love Midwesterners, and Cleveland people are so warm and fun. I had a ball living in Lakewood."

At WIXY, he fell into a dream lineup: "I loved working with Mike Reineri, Larry Morrow, Joe Finan, Bill Clark, Lou Kirby, Chuck Dunaway, [later on] Billy Bass, and Marge Bush, our fantastic music director." Knapp was said to have a mobile phone in his Jaguar XK-E for taking calls from listeners and three secretaries working around the clock at WIXY to take messages. He gave out his number on the air and the first night drew four hundred calls. Knapp said he planned to call everyone back but because of the high number would only try each number once. Boston was a major media market, of course bigger than Cleveland, and it was a nod to WIXY that it could draw such high-caliber talent.

Knapp had his work cut out for him. WIXY was sponsoring shows by some of the biggest current names in music, and that March he was assigned to report on the opening day of ticket sales for a Doors show. Jim Morrison had established himself early on as a rebel, refusing orders from staff at the *Ed Sullivan Show* to substitute tamer lyrics for the suggestive ones in "Light My Fire." Also, in a previous appearance at Musicarnival, the summer tent theater in Warrensville Heights, Morrison had opened the show with a roaring belch into the microphone. Bad-boy image aside, Jim Morrison and the Doors had a great sound and a strong fan base in the Cleveland area. Despite that, WIXY didn't want to risk its family-friendly reputation and pulled its sponsorship when Morrison was charged with lewd behavior at a concert in Miami, Florida.

Larry Morrow was also getting some high-profile assignments, among them appearing as a "singing host" for the May Company's Trend '69 fashion shows. Such shows were popular draws in the late 1960s, and the WIXY jocks even staged a "He and She" gathering at the Kon-Tiki restaurant in downtown Cleveland modeling unisex clothing with their wives and girlfriends.

To this day, people talk about the unique WIXY sound. The processing made it jump out of the speakers, and engineer Pat Richards says this was no accident. "Technically, we had Plate Reverb at the transmitter, plus Volumax and audio max. When we moved downtown a few years before, all music was put on carts, and I rode gain (watching the volume and monitoring levels) on all the songs, through an equalizer and limiter. The needle barely moved in the control room. As a result, it 'sounded louder' on the dial, even though the station had less power."

Joe Finan's popular morning show would cover any number of topics aimed at generating "water-cooler talk," drive-time subjects people discussed once they arrived at the job. Cleveland's media-savvy mayor Carl Stokes would occasionally appear on Finan's show to discuss issues guaranteed to get his message out to a huge audience. Finan also joined with music director Eric Stevens that April to write a documentary, *Sounds of Dissent,* which examined the history of various protest movements, from 1772 to 1969. But Finan heard the call of the west that April and left WIXY to return to KTLN / Denver as program director, talk show host, and part-owner. Mike Reineri replaced him at WIXY.

WIXY was quick to publicize its new morning man, and Mike Reineri's first major stunt on his return was linked to National Dairy Month. Nellie

Mike Reineri's wry sense of humor gave a new edge to WIXY's morning-drive show. (*Cleveland Press* Collection / Cleveland State University)

the cow was brought to East 9th and Euclid, one of the city's busiest intersections, and the station's Chuck Dunaway faced off against Reineri to see who could pull the most milk in three minutes. As did similar events, the promotion drew heavy interest from the media and public. The station had also taken over as cosponsor of the annual Teen Fair at Public Hall with the *Cleveland Press*. Meanwhile, on the FM dial, WMMS-FM, the city's first progressive rock station, quietly switched to a middle-of-the-road, easy-listening format.

During WIXY's heyday, radio's apparent glamour had a lot of people exploring it as a possible career, but there weren't a lot of colleges offering it as a course of study. Also, many would-be radio people really didn't want to enroll in a four-year program. Broadcasting schools had been around for some time, notably Billy Tilton's Announcer Training Center, later the Cleveland Institute of Broadcasting (CIB), at the Hippodrome Building on Euclid Avenue. A number of graduates and others who brokered time on ethnic stations like WXEN-FM and WZAK-FM to air progressive rock shows took over operations at CIB and started hiring WIXY jocks to teach for extra money. Sensing WIXY could likely be a bigger draw by sponsoring such a program on its own, on May 1, the station owners opened the WIXY School of Broadcast Technique at its Euclid Avenue building, offering the 1260 staff as instructors, along with a job placement service. The school's ads featured King Kirby at a microphone with "glamorous recording star" Dionne Warwick. Newsman Bill Clark and assistant program director Marge Bush were instrumental in running the school, which had also been accredited by the State of Ohio.

By July, within a year of the WMMS format change, Billy Bass had left that station and joined WIXY, where he was playing more progressive, so-called heavy music Sundays from 9 P.M. to 1 A.M. Nearly forty years after starting at WIXY, Bass said he realized he made a good move: "Man, did I love working Top 40! I loved the audience! Music meant a lot to me because I lived with my Aunt Jane, who worked at the Cosmopolitan Bar on East 55th and Woodland. Back in the fifties, the really good music, before Alan Freed came to town, was in the bars in the jukeboxes. I used to hang out with my Aunt Jane at the Cosmopolitan even though I . . . was just a kid. But they knew I wasn't going to be drinking. The music on the jukebox was amazing! Johnny Ray and Joe Tex . . . people like that. Music always meant a lot to me, especially R & B or real funky down-home stuff."

Bass also brought a sense of FM credibility to the WIXY job, hosting a springtime concert with groups such as Country Joe and the Fish, Tiny Alice, and Natchez Trace at Case Western Reserve University for its Committee to End the War in Vietnam and the Student Mobilization Committee. Bass also joined another founding father of the progressive FM rock movement, his old friend Martin Perlich of WCLV, and promoters Mike and Jules Belkin at John Carroll's WABU-FM later that summer to discuss coming trends for more experimental blues-based and psychedelic music.

As WIXY and its jocks were expanding their notions of music and the station's place in it, they also carried on what had become traditions. The third annual Appreciation Day was announced for July 24 and, again, it would feature a major concert lineup at Geauga Lake Amusement Park. This year, musical guests included Motown favorites Smokey Robinson and the Miracles, Tommy Roe, and a group destined to become one of the big arena draws in the 1970s—Three Dog Night. This

Mike Reineri, Larry Morrow, Chuck Dunaway, Billy Bass (kneeling), Chuck Knapp, and Lou "King" Kirby (Norman Wain Collection)

particular Appreciation Day was also designed to cross-promote the up-coming *Cleveland Press*–WIXY Teen Fair, to be held at Public Hall that August. The station's Supermen introduced the Geauga Lake crowd to some of the Miss Teen Mid-America contestants and handed out free passes to the Cleveland event.

The Teen Fair offered voting machines, which let young people voice their opinions on any number of topics, from politics to fashions, along with the 3-D Aquarius Festival light show, and music from more than fifty bands. There was also a graffiti wall, artisans' alley, plenty of cars, and a hairstyling demonstration. Go-go dancers, cosmetic demonstrations, give-away booths, and a folk music coffeehouse rounded out the attractions. The event was a major promotional coup for both newspaper and radio station. It was also an opportunity for music fans to hear some major tal-ent, as up-and-coming acts joined groups that had already made it to the national stage. The nine-day expo signed up acts like Joe South ("Games People Play"), the Peppermint Rainbow ("Will You Be Staying after Sun-day?"), Oliver ("Good Morning, Starshine"), longtime hit-maker Lou

Christie ("Lightnin' Strikes"), and The Illusion ("Did You See Her Eyes?"). Perhaps more important was the local talent getting its due, including Silk, featuring Michael Stanley; the James Gang, with Joe Walsh; Cyrus Erie, with members who would go on to form the Raspberries; the Mr. Stress Blues Band; and Damnation of Adam Blessing, who were appearing on station charts across the country. The WIXY jocks were, of course, there in full force along with newly named music director Chuck Dunaway.

Dunaway replaced Eric Stevens, who left the station to start Brilliant Sun Records with his father, Perry, a veteran music promotions man. Candace Forest, who had done wonders in her role as promotions director, also left in late summer 1969. She went on to a career in the advertising field.

But there were arrivals as well that fall, most notably industry vet Bill Sherard. He remembers,

I came to WIXY as program director in fall 1969. I had been program director of WAVZ in New Haven, Connecticut, and responded to an ad in

Chuck Dunaway left a 50,000-watt flamethrower to Captain WIXY as its program director in late 1969.(*Cleveland Press* Collection / Cleveland State University)

Billboard magazine, sending a short resume and audition tape. Norman Wain and Joe Zingale called me and repeatedly asked me if I was totally responsible for the production and sounds on the tape. After they were assured that I had solely produced the tape, they flew me to Cleveland for an interview. I recall taking my cassette recordings of various shows and productions to their space-age studios on Euclid Avenue. Unfortunately, . . . WIXY had not yet made it to the age of cassette players. I had to return to my hotel, get my portable cassette player, and return to the studios so everyone could listen to my work.

This small technical lag aside, the move to WIXY was a major step for Sherard, who was well aware of the station's promotional juggernaut. He says the staff banded together like they were fighting for their lives: "This was not unlike many AM radio stations who were feeling the pressure of FM alternative rock stations. One key reason WIXY was great is that management devoted many hours each day to developing and executing creative ideas. They not only encouraged creativity but committed the time to develop creative ideas." Sherard adds, "Norman Wain believed that WIXY's success over WKYC was due to three things: Sales, promotion, and programming. Since WIXY had an inferior signal to WKYC, and programming was generally equal, sales and promotions were the areas in which the station had to excel constantly."

Sherard came to a station that had become legendary not only in Northeast Ohio but throughout the radio industry. Such fame often comes with a good number of proportionately sized egos, but Sherard wasn't intimidated. "I had already worked in Washington, D.C., with major talents in black radio, then called 'soul radio' and had managed several morning talents in New Haven that were very strong personalities as well, so I was not intimidated. Chuck Dunaway was the most experienced talent at the time, but he was very manageable, to my way of thinking." Even so, Sherard recalled how on his first day as program director, shortly after meeting him, Dunaway "walked out of the control room, refusing to continue, since the earphone amplifier had not worked satisfactorily for some time. We were able to get him back on the air, but it set the stage for a staff of larger-than-life personalities." Sherard programmed one person less closely than the others. "Mike Reineri was another story, he says. "Norman Wain told me that they would handle Mike directly and get him to execute what programming required."

Wain says,

> We were afraid to go all talk, but we let the morning guy talk more. I'd read an article that said people who have been sleeping for eight hours have been cut off from the world for a while. They yearn for human contact again. While they want music, they also need information, and they need a discussion to wake up their brains. I told both Joe Finan and Mike Reineri they don't have the same rules as the other day parts, and they could loosen up and do something a little different, but also a lot of time and weather and what's going on around Cleveland. Finan really took off with it and built a career later on in Akron.
>
> I'm not denigrating Reineri. He just wasn't as well informed as other people, but he had a certain sardonic attitude, which we liked. He sounded like he was ticked off most of the time, which to some extent he was. Reineri was able to give the impression of a talk show without talking too much. He was really a very good talent. He became friends with Dennis Kucinich when Kucinich was an up-and-coming councilman, before he became mayor. Kucinich would drop in on the show once in a while and have other people on the phone, too. Reineri had that quality, that on-the-air sound that was just incredible and so unique. He was very good talent!

Joe Zingale added that the morning drive-time programming helped widen WIXY's listening demographic: "A lot of that was for Mom and Dad and keeping them in the loop. Talk in the morning is pretty traditional to start the day." Sherard agreed: "In my experience in programming, music has never been the key to morning ratings, even on a music-based station. We encouraged talk and attempted to produce shows that gave the [on-air] personality fodder for discussion. Sports, weather, call-ins by listeners, community events, and colorful items were encouraged. If someone mentioned something that Mike Reineri had said or done on the morning show at the office water cooler, we were successful."

The WIXY influence was inescapable. Mike Reineri was spotted on WKBF-TV (Channel 61) appealing for funds during the *Jerry Lewis Labor Day Telethon for Muscular Dystrophy,* Larry Morrow was honored at the Geauga County Fair, and all of the staff members made countless appearances. Chuck Knapp also got extra duty in September; a *Million Dollar Survey* program was added to his show, giving it an additional hour.

Perhaps inspired by the success of its School of Broadcast Technique, the station also offered a charm school at its studios that September, including lessons in makeup, posture and poise, careers, and popularity. Twenty-two-year-old London model Helen Murray was among the instructors.

In what might have been a nod to the FM mentality, which was growing among college students and those who considered themselves "heads" or "freeks," WIXY also hired a "poet in residence," Laurence Craig-Green. A poet? The station thought it could work and had Craig-Green perform his original verse in a hushed voice between songs, usually in twenty-second snippets. He came to the station after studying Eastern religions on the West Coast and having done a stint as a house poet in Venice, California. He called his work *Basic Raps.* This was a time when pop poets like Rod McKuen and Richard Brautigan were getting lots of attention. Even so, Craig-Green would not be with the station for long.

That September, Chuck Knapp was named honorary chairman of WIXY's drive to lower the voting age, which called itself the Cuyahoga County Coalition for Vote 19, with the station's offices as campaign headquarters. The amendment was placed on the Ohio ballot that November, and Knapp hosted events toward passage and reported them on his eight-to-midnight show. He also offered equal time to those opposing the issue. The issue failed, and it was a few more years before teenagers got the vote.

That October, WIXY listeners also had a rare opportunity to hear unre-leased Beatles music. The group had just released "Something" and "Come Together" from its upcoming *Abbey Road* album, another masterpiece in what was destined to become a body of incomparable work. There was a huge demand for new Beatles product, and any TV or radio show they ap-peared on would get huge numbers. To that end, WIXY played cuts from a bootleg album titled *Kum Back,* which was produced from tapes stolen from the band's sessions for its *Get Back* LP. Apple Records soon issued a cease-and-desist order, but not before WIXY scored a major on-air coup, being the first to bring these new songs to the Cleveland airwaves.

The same month, morning man Mike Reineri became part of a unique promotion, opening an Edsel dealership in the old Ohio Theater on Cleve-land's decaying Playhouse Square. He billed his "business" as "America's Largest Edsel Specialist"—an honest claim, since Edsels had been out of production since 1960. Bill Sherard, who was there, says, "We spared no effort in making the promotions sound and appear exciting. When we opened Mike Reineri Edsel, there was absolutely no listener or direct sales benefit except the 'buzz' factor in Cleveland. We purchased a 'Stop by Mike Reineri Edsel Today' jingle, had a huge client party, with a marching band and spotlights (including a full fleet of Edsels—restored of course) and gave away Mike Reineri Edsel ashtrays and wooden rulers—just like a real car dealer."

Linda Scott, who replaced Candace Forest as promotions director, also recalls that promotion and how her department morphed Reineri into "a greasy huckster." She says, "We turned that old theater into a replica of a dealership, showcasing six gorgeous vintage Edsels from all over the state and topped off the weeklong event with a grand-opening ribbon-cutting ceremony, with Mayor Carl Stokes officiating, along with live bands and dancing. It got terrific coverage on all the TV stations."

But the event wasn't without its risks, Scott remembers: "Did you ever try to maneuver a car with huge fins through a store's normal-sized front door without tearing the place apart? We couldn't get the first car in. So Norm said 'get a hack saw and cut a hole in the metal door frame. Whatever it takes, get that damn car in there!' Of course, that's what we did. I'm pretty sure the theater owner was a bit pissed off. Oh yes, he most certainly was!" No one really expected any of Reineri's cars to sell, and the promotion ran its course and closed just a few days later, the cars auctioned off for charity.

Election Day 1969 came not long after Reineri's dealership closed, and on October 29 WIXY owners Joe Zingale, Norman Wain, and Bob Weiss issued a statement on behalf of the station endorsing Cleveland mayor Carl Stokes's reelection bid. It was called a precedent-setting move, and for three days prior to the vote the station ran editorials stating Stokes was "simply the best qualified candidate for the job at the present time." Stokes was also careful about his media appearances but felt he was on friendly ground whenever he appeared on WIXY, even in the final days of what would be his successful campaign. In fact, WIXY got the mayor's only radio appearance the day before the election, due in no small part to the station's endorsement.

WIXY's music mix at the end of 1969 included everything from the Archies' "Sugar, Sugar" and Bobby Sherman's "Little Woman" to the Youngbloods' "Get Together" and everything in between, though light acts like Vicki Carr and Peggy Lee were very strong. Although the station had a very diverse playlist, it was still the top choice to host the Led Zeppelin / Grand Funk Railroad show at Cleveland Public Hall on October 24. The venue was set up with no chairs on the main floor so kids could dance or even lay prone during the show. All the WIXY deejays were on hand. Word also came down that WIXY's news voice Alexander Prescott would be heard in Grand Funk's song "Paranoid" on its upcoming LP. Music was changing, but WIXY was still the top station in town.

In fall 1969, shortly after the Beatles' *Abbey Road* album was released, rumors of Paul McCartney's death flooded the country. In what many now see as one of the most innovative music marketing campaigns of its time, WIXY's Billy Bass presented a special program examining these stories. His show featured voice lab technicians who compared new and old tapes of McCartney's voice as well as comments from Apple Record executives and even an interview with psychic Jeanne Dixon. The show wisely pointed out the folly of such rumors but still made for some dramatic radio.

That November, while WIXY was debunking McCartney death stories, there was another format change on the FM dial. WMMS decided once again to test the waters of rock radio with a syndicated, more up-tempo format called Hit Parade, which posed no real challenge to WIXY because so few cars had FM tuners. (Those that did were usually in upscale cars driven by older motorists who tended to drift toward mellower offerings like Joe Black's "Journey into Melody.") WMMS's new format ran the gamut from Tony Bennett and Henry Mancini to the Beatles and Tom Jones, but only middle-of-the-road offerings from those acts.

Thanksgiving was right around the corner when WIXY sent Larry Morrow to combat zones in Vietnam armed with a tape recorder. Bill Sherard tells the story: "We sent Larry Morrow to Vietnam to interview Cleveland-area hometown soldiers [and tape] Christmas greetings for their families back home. We listed their names and air times in the *Plain Dealer* and used the slogan 'WIXY brings the boys home for Christmas.'"

Morrow was in 'Nam for two weeks and would always be affected by the response he got from those men in uniform. Recalling those days to an interviewer in the 1970s, he said he went to Southeast Asia for many important reasons.

I was not only sent on behalf of WIXY to interview all the Cleveland area boys, and all the Pittsburgh boys for our station in McKeesport. I was to bring back the interviews for Thanksgiving so the parents and friends could hear them over the air, but also to judge the morale on behalf of the Pentagon of all the troops in Vietnam . . . the enlisted men and the officers. It was a frightening experience because I was there three weeks and it seemed like every base that we approached we were greeted with a red alert . . . , and a red alert at that time meant that the base was under attack! We literally dodged bombs! . . . When we went to Cam Ranh Bay, which is the halfway point between North and South Vietnam . . . the demilitarized zone . . . that is the biggest military hospital in the world, and the first area I went through was the amputee section. It was frightening, and I could hardly wait to get out of Vietnam.

Morale at the time, and remember we're going back to 1969, was very good among the enlisted men. Most of the enlisted men were kids 18 to 20 years old. They loved being in Vietnam and fighting for their country. They also sympathized with the South Vietnamese people, who . . . were in a quandary because they didn't know whether they should be faithful to the Americans there because they worked on our bases during the day. They also were unsure if they should work for the North Vietnamese at night, because the Viet Cong threatened harm if they didn't sympathize.

It was a profound experience for Morrow. "It was amazing. You're halfway around the world in a war zone, and you go up to a kid and say . . . for example . . . 'Larry Wilkins? I'm Larry Morrow from WIXY,' or he'd see my WIXY 1260 patch, and he'd say, 'Hey, Duker! What are you doing over here?' A lot of people knew me from the time they were high school kids, and generally they were all very nice. It was quite an experience for me. At that point that was the highlight of my career."

The interviews were full of emotional messages for family and friends back in Northeast Ohio. For example, Dan Weinar of Cleveland Heights mentioned he "graduated from West Tech High School and wanted to wish a Happy Thanksgiving to his wife, Nora, and her parents," and he added a few words of appreciation for WIXY, "You know, Duker. You got the best station in Cleveland. In fact, it's the only station in Cleveland!" PFC Ross Jacobsen of Canton, a graduate of Canton McKinley High, wished happy holidays to his mom and dad, and gave a dedication to his girlfriend, Eleanor, saying he expected to be home by Christmas—or hoped to. Leroy Creesen Jr., a Glenville grad, passed on to his cousin Mrs. Eleanor Richardson that he was doing fine and life wasn't too bad for him in 'Nam. He worked as a supply clerk in Da Nang and was looking forward to his return to the States in June 1970. Charles Snowden III told his dad to enjoy the holiday and not worry—he would be home soon. He added that his dad was the "grooviest guy" in the world, and recalled with fondness playing football with him.

There was also an odd sidebar to the Cleveland-Vietnam link. Sherard tells the story: "I do remember Mike Reineri covering the family in upstate Michigan whose son died in Vietnam, and among his returned items was a class ring. Shortly after the grieving family received it, it began growing a 'hair,' a filament that kept extending, longer and longer. Mike interviewed them and flew up there to see the thing and report back in person."

Promotions director Linda Scott accompanied Reineri: "Messages were supposedly sent through spectral threads emanating from his class ring," she recalls. "One weird weekend I assure you, but it made riveting material for Mike's listeners." Upon his return, Reineri confessed to believing the story.

While many WIXY listeners were thinking of their loved ones in Vietnam, they perhaps welcomed distractions like the annual WIXY–May Company Christmas Parade. This year, it kicked off on November 29 with the up-and-coming Osmond Brothers sharing the grand marshal duties. As expected, the streets of Cleveland were jammed with onlookers hoping to spot special guests like Miss Ohio Kathy Lynn Bauman and actress Kathy Garver from the hit CBS-TV series *Family Affair,* as well as music from the Damnation of Adam Blessing and Jay and the Americans. TV8's Big Chuck and Hoolihan were on hand as well. And thousands of parade-goers stayed on after the festivities, to do some shopping and most of all get warm in the downtown stores.

But distractions like the parade and other holiday cheer didn't keep the war from people's minds for long. That December, WIXY joined other news organizations across the country to air the Selective Service draft lottery. Vietnam had been a dividing issue in the generation gap, and was a sore point between the youth of America and their parents, who had answered the nation's call during World War II.

Also, while WIXY had celebrated a banner year, things weren't smooth behind the scenes. Engineer Pat Richards recalls, "There almost was a strike in '69 but the jocks did not want to support it, so NABET [the union] just went along with whatever the company wanted to give."

1970

The entertainment world had changed drastically since WIXY first went on the air, with the Top 40 sounds of the British Invasion and Motown giving way to psychedelia and Woodstock. But the WIXY music mix that year was a wild combination of styles and songs. Jocks played the Top 100 songs of 1969 on New Year's Eve and the first day of January 1970, with the Archies' "Sugar, Sugar" beating out the Rolling Stones' "Honky Tonk Woman" for the No. 1 spot. Just a week later, Mike Reineri offered $25 to the first female listener to brave the winter winds and snow and show up at the WIXY studios in a bikini. This was a lot of money in 1970; eleven minutes after he made the announcement an eighteen-year-old listener knocked on the studio door in a two-piece bathing suit.

Reineri showed some skin as well that January. The station erected a billboard that parodied its painted woman promotion—with an un-clothed Reineri looking over his shoulder. It announced, "Reineri's Back," and didn't get anything near the favorable response the young woman had. This sign was quickly taken down. Another cold-weather promo-tion took place just two days after the bikini call at the East 9th Street pier; listeners were asked to come out and cheer on Chuck Dunaway and Larry Morrow in the first annual Frost Bite Row Boat Race on Lake Erie. Billy Bass got into the cold weather promotions as well, hosting a Teens Go Mod maxi-coat fashion show for the national March of Dimes.

Larry Morrow's musical talents were also coming to prominence, with "Pammy Jo," a song he'd written for local singer Jerry Tiffe on Scepter Records. Unfortunately, because of Morrow's participation, WIXY could only play the song a limited number of times, and no other station was likely to pick up a tune benefiting another disc jockey.

The weekly WIXY survey was a popular promotion, and the annual songs-of-the-year presentation was a huge draw as well. Perhaps trying to capitalize on that popularity in the cold winter months, in mid-January the four-year-old station issued a list of what it considered the Top 100 "milestone" records of the previous decade. Bill Sherard, station manager Norman Wain, and deejays Chuck Knapp and Chuck Dunaway compiled the list. The key word here was "milestone"; songs didn't receive their placing based on popularity. As Sherard told the *Cleveland Press,* the committee chose based on how well songs reflected their particular era. That might explain why songs like "MacArthur Park" by Richard Harris and "Ballad of the Green Berets" by Sergeant Barry Sadler placed higher than several tunes by the Beatles or Stones, but not why the Association's "Cherish" would be considered the top song of the 1960s.

In early 1970, the station lost one of its pioneers when engineer Ralph Quay died. He'd been with the WDOK staff since 1951 and had made the transition to WIXY with its new ownership. He was just fifty-three years old when he had a heart attack at the WIXY transmitter site in Seven Hills. He died a short time later at Parma Community Hospital.

No matter what changes came to WIXY, promotions were a constant. It wasn't easy to keep coming up with innovative campaigns and contests, but WIXY more than held its own in that department. Lou Kirby offered one that January called "How Far Can a Guy Get on a Shoestring?" It started with Kirby trading a shoestring for four bottles of aftershave. Each day he would trade the spoils of the previous transaction for something offered by listeners. Kirby's hope—and the station's, as well—was that when the contest ended, he might have something really valuable he could donate to charity.

By April 1970, the WIXY 60 survey had been trimmed to between forty-eight and fifty-four songs as a means of streamlining the format so more of the list could be played on a regular basis. On April 12, a half hour before the Indians-Senators matchup, the WIXY Supermen softball team got a chance to play at Cleveland Municipal Stadium against a team of Playboy Bunnies, led by the magazine's July 1969 centerfold, Nancy McNeil. As loyal as WIXY fans were, it's not likely the men in the crowd were cheering for the boys from 1260. As if the WIXY-Bunny game wasn't exciting enough, the station gave away a new Pontiac Firebird at that game as grand prize in its Cartune Contest.

Progressive rock had been around for years, and Glass Harp, a power trio out of Youngstown, was getting a great deal of attention. Riding this

wave, Bill Sherard formed Idyllic Productions with the idea of getting
the Harp a major recording contract. The band was eventually signed,
and in February 1972, Idyllic Productions helped make local television
history when it presented the band in a live concert from the WVIZ studios, with a stereo simulcast on WMMS-FM.

The threat of FM rock was starting to loom larger, and WIXY wanted
to keep its position as top dog on the AM dial. Sherard remembers this
time well. "There was very much pressure to keep the momentum going," he recalls, "even though WKYC had already dropped their Top 40
format when I arrived, the pressure seemed very intense compared to
my previous assignments. Underground radio was beginning to have an
effect, with record labels selecting particular cuts from hot AOR [album-
oriented rock] albums and promoting them as singles." WIXY wasn't
about to change its obviously successful format, so again it relied on its
innovative promotions team. One result of the promotions push was a
competition based on its very popular "Francine" contest. On April 24,
the station had a contest on the corner of East 9th and Euclid—its First
and Last Annual Raquel Welch Look-Alike Contest, offering fashion ac-
cessories as an incentive to contestants.

WIXY's Larry Morrow was an amateur chef, and by all accounts a good
one. One of the hallmarks of his midday show was a segment called "What's
Cookin'?" in which he would give an ingredient at a time throughout the
show until the audience had an entire recipe to copy down. Morrow's spe-
cialty was Arabic and Syrian cooking; with Northeast Ohio's ethnic mix
he may well be credited with introducing stuffed grape leaves and baked
kibbee into homes more accustomed to kielbasa or cannoli.

The big surprise in radio circles in early June was the announcement
that Lou "King" Kirby would be leaving WIXY, to be replaced by "Big John"
Roberts out of WIFE / Indianapolis. The now twenty-eight-year-old Kirby
wanted to spread his wings a bit and do more TV work, though he was
among those said to be looking into a position at 50,000-watt WGAR.

After Kirby left, of course, many WIXY diehards remained. Billy Bass,
for example, brought an FM sensibility to the AM airwaves of WIXY on
several different levels. He urged support for the Cleveland Free Clinic
and hosted shows like *Cleveland Drug Crisis,* which pointed out that drug
abuse was a problem not restricted to young people. Bass also cohosted
the Who concert at Public Hall that June with Larry Morrow, who flew in

with the group from the Cincinnati show. By this time, Cleveland was seeing at least one rock concert just about every weekend, and Belkin Productions featured the station's call letters in its weekly rundown of shows.

That summer's chief competition came from stations on the FM dial.
WGAR-FM had changed its call letters to WNCR to reflect its owners,
Nationwide Communications. On July 6, the new station went on the air
with program director Jerry Dean saying it wanted to reach the college-
age crowd and would play "70 to 75 percent of the hits heard on WIXY.
But we won't play any bubblegum music, no sloppy lyrics, mostly message songs." WGAR had a big job ahead of it, as FM traditionally appealed to a much older audience. Still, options were now becoming more
plentiful for the audience that didn't care to hear a variety of songs that
included "Gimme Dat Ding," by the Pipkins, or Robin McNamara's "Lay
a Little Lovin' on Me," which all appeared on the WIXY lists. However,
WIXY still had the respected Billy Bass on its staff, and it kept him very
busy with highly visible promotions and appearances.

Still, when it faced the lush sounds of stereo FM, the writing seemed to
be on the wall for AM's mono signal. "WIXY was programmed for ratings,
but sales drove the ship," says Sherard. "As Bill Drake's successful 'more
music' format swept the country, playing songs back to back without DJ
personality chatter became a priority." He also described another key distinction between WIXY and stations like WGAR: "Another difference which
made fighting the battle difficult was the increasing schism between those
who were 'on the bus' (pot smokers) and those who weren't. Top 40 stations' deejays were still appearing at public events in radio-station blazers
while AOR stations jocks wore jeans, T-shirts, and leather jackets."

On a deeper level, he reports,

> the Top 40 format parameters were becoming stretched. . . . Having to
> program Tony Orlando's "Tie a Yellow Ribbon" followed by Led Zep-
> pelin's "Stairway to Heaven" is a good example . . . [of] a train wreck.
> Granted, that was originally the beauty of Top 40, its ability to com-
> bine different musical genres as long as they were widely popular. I
> recall WABC playing Louie Armstrong back to back with the Beatles.
> WIXY was very slow to adjust to this. . . . The "WIXY Triple Play" was
> little defense against FM stations who played music sweeps in almost all
> quarter hours, but it typified Top 40's hesitance to reduce commercial
> spot loads and instantly lose revenue. In retrospect, if broadcasters had

made a commitment to dramatically reducing the spot loads, it may not have done much to slow the erosion to FM in the long haul, and it would have reduced revenues immediately.

After hearing the stories from 'Nam, WIXY's creative team decided to bring a little cheer to those who returned injured from the war zone. Linda Scott, who helped pull the program together, remembers, "The Vietnam War was raging and many of our young kids came back maimed and recovering in the Vets hospital, unable to attend any of our station sponsored concerts. So I came up with the idea of putting together a talent show just for these vets. We held auditions and had an unbelievable turnout. Once the word got out, covered by both papers, . . . many of the local professional groups contacted us and wanted in." Once the lineup was chosen, Scott says, the event "was held at the Veterans Administration Hospital . . . on August 5, 1970. We called it the Salute to the Veterans Show. A couple of the local bands that played were Wild Butter and Knights of Brass. I have to say, we put on a fabulous show, and this was one of WIXY's finest hours, as far as doing a solid good for the community."

The WIXY jocks joined a Cleveland legend in mid-August, when they appeared in a special on WEWS hosted by Ernie Anderson, who reprised his "Ghoulardi" character for a final time on local airwaves. The program also featured comic actor Chuck McCann, and the WIXY jocks were invited to take part in the show and promote their appearance on it. The show featured some comedy sketches interspersed between old movies, much like the old Ghoulardi shows on WJW. 1260 also staged one of its weirdest promotions ever that month, asking listeners to call in and deliver the "cry of the wild mongoose." The prize was a trip to Duluth, Minnesota, to see a stuffed mongoose, said to be the only one in the United States. The station followed that one with a contest in which it played *Sesame Street*'s "Rubber Duckie" every hour and asked listeners to send in postcards describing what they did with their duckies in the tub. Some wondered if the station was simply running out of ideas.

By summer 1970, the onetime WIXY 60 survey had trimmed its list to thirty-eight songs. That's not to say the station was slipping: it was a solid No. 1 or 2 in morning drive through the year, with Mike Reineri pulling in as much as 17 percent of the audience in that very competitive time slot. Larry Morrow's middays showed solid numbers, usually No. 1, and afternoon drive to midnight was a lock for the top position—usually doubling the audience of its second place competitor. Perhaps because of WIXY's dominance, WGAR started trading on-air jabs with rival WHK, mentioning call letters, but wisely staying away from the No. 1 station.

By September, the WMMS experiment to draw ratings with its Hit Parade format was drawing to a close. It had no excitement, and the FM listener had no incentive to tune it in. To that end, 'MMS management decided to go back to a live format and hired two former WIXY jocks, Lou Kirby and Dick Kemp, as the cornerstone of the sound. Kirby had been off the air for three months, and Kemp had apparently had enough of McKeesport. But as soon as their hirings were announced, the WIXY attorneys made their move pointing out the two had noncompete contracts barring them from working for another Cleveland station for a set period of time. The fight went down to the very last minute, but the two stations finally reached an agreement for Kemp and Kirby to debut on WMMS. Meanwhile, Billy Bass, one of the vets from WMMS's first stab at progressive rock, was now on WIXY's prime-time 6 to 10 P.M. shift.

Also that month, WGAR premiered its new format, with Jack Thayer of KXOA / Sacramento taking over the reins as general manager. He brought along two other heavy hitters from that station: John Lund as program director and Don Imus as morning deejay. The new WGAR hoped to sell its mix of softer sounding current hits and oldies dating back to 1955 to the same eighteen-to-thirty-five audience WIXY had dominated. While the new 1220 was not an all-out Top 40 station, its selection of oldies had been pop-chart hits and the air personalities and jingles were directed at the Top 40 market. WGAR also muddied the waters by running a newspaper ad that commanded: "Mike Reineri, call WGAR . . . but please turn your radio down!" No one was exactly sure what it was supposed to mean, and Reineri ended up fielding phone calls from listeners wondering if he was preparing to jump ship. WGAR-FM, now called WNCR, played many of the songs on the WIXY charts; station management thought there could be a crossover of fans hoping to hear that music in stereo along with more in-depth coverage of the personalities who made the music.

WIXY's Norman Wain couldn't easily dismiss that one-two punch. He later said, "WGAR sure was competition. You see, things change. The WIXY thing was so hot and so intense. It's hard to sustain something at a hundred miles an hour all the time. We worked hard at it, and we kept doing what we did best, but at the same time we knew that the era was going to have some problems with the coming threat of FM radio."

That October, WIXY showed why it remained the top radio station in town when it aired an innovative series over two weekends, featuring the music and airchecks from some of radio's biggest personalities. It was a rock-radio history lesson, with clips of greats like Alan Freed and B. Mitchell Reed. Its playlist included everything from Led Zeppelin to the Carpenters, but the FM dial continued to threaten, as stations like WNCR and WMMS were getting more press.

WIXY continued to endorse political candidates that October, coming out in favor of Howard Metzenbaum's Senate bid. Also that month, WIXY sponsored a charity "dig in," with each participant donating a dollar an hour to shovel dirt at specified locations in search of prizes. They were good prizes, too, like a brand-new Chevy Vega, cameras, and a seven-day trip to Mexico. But Metromedia, the corporate machine behind WMMS, was serious about its new attempt to win listeners, placing large ads in papers and magazines asking, "Has rock music gotten too freaky for radio?" and showing a wild-haired youth with a guitar.

The WIXY station still had a lot of clout, though, as anyone who turned on a TV set in late October could see. The station's Larry Morrow hosted the Miss Teenage Cleveland Pageant on WUAB (Channel 43); the contest's winner went on to the national Miss Teenage America contest. His cohost was actress Susan Strasberg.

Radio wars were nothing new, but a promotion in November 1970 promised to take them to new levels. The *Cleveland Press* ran a contest to determine Northeast Ohio's favorite disc jockey. As many as fifty-eight personalities at eighteen stations were in contention for votes, and WIXY's Billy Bass was a leader at the start followed by Ted Hallaman at WHK, Mike Payne at WJMO-AM, Jack Reynolds at WHK, and Jim Runyon from WKYC. The contest was a smart move for the *Press* because the jocks talked it up, which meant a lot of free advertising, not to mention newsstand sales, as only official votes cast on *Press* ballots would be counted.

It was a very tight race going into the final days, with Runyon and WJW-AM's Ed. Fisher neck and neck, though WGAR's Norm N. Nite and WERE's Howie Lund showed considerable strength. When the smoke had cleared, in the final vote WIXY's midday man, Larry Morrow, was Cleveland's Favorite Disc Jockey. He was followed by Runyon, Fisher, Nite, Reynolds, Hallaman, Lund, Al James, Payne, and WKYC's Larry Kenney.

Morrow commenting some years later on that honor, "I think it was because the people in this city had the opportunity to know me personally. There were other people equally as popular and talented . . . some more popular . . . but at the same time, they just didn't have the opportunity to know these people personally. The people that knew me went out and voted."

WIXY's annual May Company Christmas Parade was also held at the end of the year, with an all-star cast that guaranteed additional publicity. The 1970 parade showcased David Cassidy, Susan Dey, and Danny Bonaduce from TV's *Partridge Family* as grand marshals; singer Ronnie Dyson; and 1970's Miss Ohio, Grace Elaine Bird. This year, the parade was estimated to draw as many as two hundred thousand people to downtown Cleveland! Of course, it also prominently featured the WIXY Supermen, including the latest addition, Ron "Ugly" Thompson, who wore a paper bag over his head. But that lineup was in for a major change.

On December 17, WIXY announced that program director Bill Sherard would be leaving the station, for a position as WNCR's morning-drive jock. Billy Bass also turned in his resignation, effective December 28, when he would join WNCR-FM as program director and air personality. Bass's replacement, at least for the time being, was another longtime Cleveland favorite, Chuck Dunaway. Newcomer Ron Thompson didn't last long at WIXY, after a forbidden word sneaked its way onto the air.

Later that month there was another, much bigger bombshell. The latest Arbitron ratings showed that WERE knocked WIXY from its No. 1 position. 1260 came in second, and WGAR now attracted a lot more younger listeners. WIXY reportedly lost 84,000 listeners since the last ratings count, but it still showed a very impressive average of 386,600 total audience members. WGAR, WNCR, and WMMS were apparently drawing away some of WIXY's young listeners.

FM was making small but obvious gains. But perhaps the bigger reason for WIXY's falling ratings was its wide mix of music; progressive rock fans would have to sit through Perry Como and Tony Orlando and Dawn to get to songs by Neil Young and Santana. On the FM side, it was one-stop shopping. Columnist Bill Barrett noted a lot of crossover on other stations: Of the thirty-eight songs now being published as part of WIXY's current survey, eleven of them could be heard on WJW-AM, which boasted a much softer sound. That was evident in WIXY's year-end survey with Norman Greenbaum's "Spirit in the Sky" taking the No. 1 spot, followed by "We've Only Just Begun" by the Carpenters.

1971

By January 1971, it was evident that WNCR was in it to win; as part of the plan, the station picked players from the WIXY bench. Bill Sherard took over WNCR mornings; Ron Thompson got the midday show with his wife, Kaye, who had also worked at WIXY; and Billy Bass did evenings, 6 to 9 P.M. But 1260 was still going strong on high-profile events like an annual bridal fair and sponsoring major rock concerts that drew thousands of people at a time.

The rock musical *Hair* had been getting headlines ever since its 1967 debut on Broadway—partially because of a controversial nude scene, but it also boasted groundbreaking music. When the play went on tour, WIXY sponsored the opening at the Hanna Theater on Playhouse Square on March 9 and staged an Appreciation Night to thank national media for support and encouragement. Before the curtain rose, the industry reps joined at the Pirate's Cove club in the Flats for a preshow party, which was even attended by the cast! The station transported party-goers back and forth by bus, and everyone got into the act, with then WIXY sales rep Tony Volardo also providing music with his band. Larry Morrow would recall years later, "It was unusual to see people like Mike Reineri and the rest of the jocks in tuxedoes!" WIXY also generated plenty of promotional opportunities linked to the play. Former promotions director Candace Forest recalled an earlier event that drew a lot of attention. "I do remember us finding Cleveland's 'Hairiest' person. It was a contest to win a trip to New York to see *Hair* on Broadway. I remember it because I got to go as the 'official' host from the station. It was a lot of FUN—that word just keeps coming up doesn't it?"

The race for ratings is often a volatile one, and despite its bump from the top in autumn 1970's Arbitron book, the February Pulse ratings survey showed WIXY shooting right back to the No. 1 position, with an estimated total audience of 423,000, followed by WERE, WHK, and WGAR. The only FM station to place in the top ten overall was WIXY's soft-sounding sister station, WDOK.

The City of Cleveland was celebrating its 175th birthday that year. The Sesquicentennial Committee partnered with WIXY to crown Miss Cleveland Super Sesqui, who would act as the city's official greeter. She would receive a complete May Company wardrobe, use of a new Ford, and other prizes. By this time, the city was losing population to the suburbs and shutting down after the business day, but it hoped to inject some excitement into its celebration and life into its after-hours downtown by linking with its top radio station. Among the station's many connections with the event was Mike Reineri's Raiders, a volunteer group that cleaned up Public Square and other locations for the city's yearlong anniversary.

At the same time WIXY sponsored the controversial band Alice Cooper at Cyrus Erie West, a teen dance club where the station would occasionally sponsor events or acts, even holding a press conference upon the group's arrival. While Cleveland was abuzz with its grand birthday celebration and wild concerts, there was a lot happening at WIXY under the creative supervision of assistant program director Marge Bush and program director Chuck Dunaway.

A major change was also in the works that would shake the local radio scene that spring. On April 23, 1971, Globetrotter Communications of Chicago—which owned the basketball team of the same name as well as Chicago's WVON radio—announced it had reached an agreement to merge with WIXY and WDOK's parent company, the Westchester Corporation. That deal included the WIXY School of Broadcast Technique but not WIXZ / McKeesport. Needless to say, there was great speculation how it might affect local radio.

The bottom line was that WIXY would no longer be owned by Clevelanders. It seemed inevitable. Norman Wain explained: "We decided that if somebody came along with a great price, we would go. Sure enough. Somebody came along, and we went. Like idiots, we didn't hang on to WDOK. We should have but couldn't. They wanted to buy both AM and FM; FM was just beginning back then, but we could see the handwriting on the wall." Joe Zingale would later say that Globetrotter had considered a number of FM properties but they decided to concentrate on Cleveland. The merger still needed government approval for a deal that would give Wain, Zingale, and Weiss shares of Globetrotter while giving the Chicago company control of WIXY. Before the deal, WIXY and WDOK were among the last of the locally owned radio stations, but now, only WZAK-FM remained in local hands.

That May, while the station sale was still just talk, WIXY drew on its strength in drawing crowds for major promotions by sponsoring a hot pants contest in conjunction with the clothing store chain with that same name. This one took place at Public Square on a Friday and again drew a huge audience to downtown Cleveland to enjoy a chilly but still "hot" lunchtime in the city. It drew wide media attention, with more than fifteen hundred people—mostly men—gathering in front of the William Ganson Rose band shell to see as many as two hundred women compete for the Miss Hot Pants of Cleveland title. From a stage plastered with signs bearing slogans like "When you're hot, you're hot," a WIXY jock cried out, "How we doin' out there in hot pants land?" and was greeted with a thunderous response. Police had to clear dozens of spectators off bus shelter roofs, with some even climbing into trees. Amateur cameramen scurrying for every possible shot trampled flower beds.

Each of the contestants affiliated herself with a Cleveland-area company or organization, among them WKBF-TV, the Osborn Medical Building, and even the city of Lakewood. When the smoke had cleared, a seventeen-year-old part-time model from Valley Forge High School, Shirley Malin of Parma Heights, had taken the title, and Euclid's Sandy Reckner, a twenty-four-year-old employee of Mr. Angelo's wig shop, came in second. Third place went to another Euclid resident, twenty-two-year-old Donna Thomas, who was employed by Coca-Cola. The three blondes all qualified for the national championship, to be held later that month. WIXY also paid a sizable fee to the Cleveland Properties Department to clean up after that event.

May was also the month for the station's annual Appreciation Day, and the 1971 concert featured Woodstock favorites Sha Na Na; the Buoys, of "Timothy" fame; and up-and-comers Delaney and Bonnie and Friends. Public Hall was hopping that weekend. The free WIXY show started at 11 A.M., with Frank Zappa and the Mothers of Invention headlining a show there that night, supported by Humble Pie and Head Over Heels.

The lineup continued to change at 1260; that September it included Mike Reineri on mornings, Larry Morrow continuing his midday dominance, Chuck Dunaway holding down the fort from 2 to 4 P.M., former CKLW screamer Steve Hunter until 8, Chip Hobart to midnight, and

Bobby Knight entertaining until 6 A.M. WIXY saw a lot of turnover. Some names, like Jim Conlee, earlier in the year, and even Hunter lasted only a very short time. And Chuck Knapp had left for Atlanta's WQXI at the beginning of the month. WIXY's strength was also its curse; many of its staff left for bigger markets based on the strength of the station's call letters on their resumes.

The station's change in ownership played a part in some of the staff members' departures. Of his own leaving, Knapp later explained, "I stayed until Norman Wain sold the station to the Globetrotters. They play basketball; they don't do radio. I went from WIXY to WQXI in Dixie [wicksee to quicksee], after Globetrotters refused to put me in their starting lineup with Meadowlark Lemon! (A joke! A joke!) I was on the Black Economic Union team with Jim Brown and Paul Warfield. I was their 'token white guy.'"

The sale had been in the works for several months, and Cleveland's radio scene was already beginning to feel its effects. That October, as many WIXY staff members were leaving the station, the Federal Communications Commission approved Globetrotter Communication's merger with the Westchester Corporation for a reported sale price of $14.5 million. The next month, former WLS / Chicago head Gene Taylor came in to run WIXY, replacing the trio who made Cleveland radio history.

Even though 1260 had a major changing of the guard, many long-standing traditions continued. The WIXY–May Company Christmas Parade kicked off that November with yet another all-star bill that included actor James Darren; Glass Harp; TV's kid rockers the Banana Splits; Miss Ohio Judy Ann Jones; and, not surprisingly, Meadowlark Lemon of the Harlem Globetrotters.

WIXY's creative spirit also remained strong, despite all of the changes. In December, the station aired a special presentation by the "WIXY Players," which included Marge Bush, Chuck Dunaway, Mike Reineri, and other members of the staff in a parody of the morality record "Once You Understand" by Think. The WIXY satire featured a father urging his son to forget about his violin and to grow his hair long and join a rock group, as well as giving the lad as much money as he wants. The family's mother urges their daughter to stay out late, but the kids want a normal

Facing page: WIXY never failed to bring out hordes of young men to promotions like the Miss Hot Pants of Cleveland pageant. (*Cleveland Press* Collection / Cleveland State University)

life. It ends with the father breaking down in tears when he hears his son won a four-year scholarship to Harvard.

In December, Jeff McKee also arrived at 1260. A high school dropout who had worked at six stations in four years, McKee was also a natural on the air and was only nineteen, not even legally old enough to drink when 1260 hired him. It was a year of transition for WIXY 1260, and promised to be as much of a roller coaster in the months that lay ahead.

1972

Early 1972 saw former WIXY jock Dick Kemp announce his retirement; he swapped radio for a farm in Middlefield and a career as an antique dealer. But Kemp couldn't permanently retire from radio; in 1972 he took a job at WOBL / Oberlin. Meanwhile, in January of the same year, Kemp's old station WIXY again aired the draft lottery numbers. On a lighter note, in the same month it offered a handout at record stores, *Phonograph Record Magazine*. It was a nationally produced publication, a promotional paper similar to *Scene* or even *Rolling Stone*—with a middle insert featuring WIXY news. The station was still one of the most important radio outlets in the country, host to many of the most popular artists in the country. For example, program director Chuck Dunaway often hosted recording artists like Peter Yarrow, of Peter, Paul & Mary fame, at informal gatherings in his apartment.

In another patented 1260 move, the station launched a wild campaign, its Keep on Truckin' with WIXY contest, which offered a camper as a prize. It was based on onetime Cleveland artist Robert Crumb's famous underground comic. There were plenty of other prizes, too, and to win, a listener had to answer the phone or door by saying, "Keep on Truckin' with WIXY" or even put signs on their lawn with the slogan. The station promised to give away a total of $25,000 in prizes in that contest.

But in the midst of the contest excitement, WIXY was not prepared for a major announcement from one of its franchise players. That May, Larry Morrow announced he would be leaving the station effective June 30 to devote his full time to his jingle company, Morrow's Music Machine, and writing and producing tunes for commercials; he'd already done jingles for clients ranging from Smucker's to B. F. Goodrich. He promised to honor the station's noncompete clause if approached by another radio outlet before the end of six months.

Major changes were on the horizon when popular midday man Larry Morrow announced his departure. (*Cleveland Press* Collection / Cleveland State University)

While WIXY was getting used to its new ownership, the Cleveland radio scene was shifting on other fronts. A controversial new format at WERE, spearheaded by morning man Gary Dee, was drastically changing the radio scene. By June, Dee and the station's brand-new "People Power" format had skyrocketed to the top of the ratings. WERE's air sound—now loud, flamboyant, and controversial talk—was said to be attracting some younger listeners as well. WERE bumped WIXY to No. 2, but its new format's true test would be in sustaining those numbers against the longtime favorite at 1260. Another key element to WERE's resurgence, though, was the similarity in playlists at the music stations. As *Cleveland Press* columnist Bill Barrett pointed out, WIXY, WGAR, WKYC, and even easy-listening WJW were playing many of the same songs.

WIXY debuted its new midday man on July 10. Bob Shannon came from KJR / Seattle to face the tremendous job of filling Larry Morrow's shoes. Shannon was no stranger to Cleveland, having worked at WKYC from July 1968 to early '69. He was a veteran of the radio wars who had seen a lot in his twenty-nine years. And he was more than willing to tease the competition at WKYC: when he was introduced in a *Cleveland*

Press article, Shannon recalled WKYC as having a hard time keeping talent, going through twenty-two jocks in two years. He said, "I had senior-
ity when I'd been there only two months!"

He also took a crack at the WKYC audience: "One Saturday morning
the engineer came running into the studio at WKYC while I was on the air.
The *David Brinkley Journal* tape from New York was lost. I recorded two
minutes using my best Brinkley voice and said nothing of any content. To
show you how many people were listening, nobody noticed it." To be fair
to WKYC, most people in its target demographic tried to sleep in on weekends after late Fridays, and if the impersonation was that good, maybe
they didn't have a reason to suspect it was anything but legitimate.

Shannon also turned down jobs in bigger markets like San Francisco to
return to Northeast Ohio because he really loved Cleveland. His job now
was to see if Cleveland returned that love. WIXY's management decided
that Shannon would be officially introduced at WIXY's Spirit of '72 Appreciation Day, held July 9 at Edgewater Park. These were the days of
mega-rock shows, riding the success of the country's huge rock festivals.

WIXY's '72 Appreciation Day drew thousands to hear local and national recording
artists. WIXY's Jeff McKee and Chuck Dunaway with singer Buffy Sainte-Marie.
(*Cleveland Press* Collection / Cleveland State University)

Edgewater had hosted similar events sponsored by the FM rockers and, taking a page from the Woodstock diary, WIXY booked Country Joe and the Fish as one of its headliners. Given Joe's "Fish Cheer" at Woodstock ("Gimme an F! Gimme a U! . . ."), it was obviously a pretty brave move by the folks at 1260. The rest of the bill included folk singer Buffy Sainte-Marie; Jerry Garcia protégés the New Riders of the Purple Sage; the hard-rocking Brownsville Station; Lobo; Tony Joe White; and three groups with Cleveland roots: the Raspberries, the James Gang, and Brewer and Shipley of "One Toke over the Line" fame. WIXY provided free shuttle buses to Edgewater from the municipal parking lots downtown, and there were medical facilities on site as well. The promoters, Belkin Productions, built two stages so equipment could be set up on one during a performance on the other. Mayor Ralph Perk's office also issued a public service award to morning man Mike Reineri and his Raiders for helping clean up the city. The group would likely have its hands full after the daylong concert on the shores of Lake Erie.

While WIXY continued to celebrate with grand music events, its air talent continued to come and go. Terry Stevens arrived that summer from KFMX / Omaha to do afternoon drive and replace Matt Quinn, who had made little impact during his short stay at the station (less than a year). Stevens had started out as a newsman in South Dakota and admitted it would be a challenge to make an impression in a market the size of Cleveland. There was growing competition on the FM dial as well; WGCL, for instance, announced it would switch from an automated format to live disc jockeys. Plus, inexpensive FM converters could now be installed in cars, offering even more automotive listening variety. To battle that increasing challenge, WIXY decided to concentrate on what it did best. While its format caused some restriction, promotions were wide open at WIXY. To that end, disc jockey Mike Kelly underwent a complete physical at the Cleveland Clinic to prepare for an attempt to break the world's Ferris-wheel riding record, which Pogo Pogue of Honolulu had set in 1966. Kelly would be sitting on Cedar Point's Ferris wheel for a long time: the stunt would require him to be in the gondola from at least 10 A.M. to 11 P.M., even sleeping on board in a specially fitted cot, for at least seventeen days. Of course, listeners were invited to come out and cheer Kelly on and see if they could catch him failing in the attempt. Needless to say, Kelly broke the record and made it to the Guinness book, and he was feted with champagne, a trophy, and a week's vacation. But not everyone saw the value of the stunt. *Esquire* magazine would later give Kelly its Dubious Distinction award for the promotion.

Even though radio was well into its transition to the FM dial, there was still a battle in the morning-drive slot, and AM was reigning king. The latest Pulse ratings showed WERE continuing its fast rise, though WIXY's Mike Reineri more than held his own against WERE's Gary Dee, WGAR's John Lanigan, and WJW's Ed. Fisher. Pulse and Arbitron sometimes showed different ratings for the same time periods, and those numbers could be crunched to just about anyone's advantage.

By November, there were reports that Norman Wain had an office at WIXY but was rarely there. Regarding these reports, however, Wain says,

> Let me tell you what happened. George Gillette was an incredible negotiator, and he knew how to buy and sell stuff. He came out of the investment banking business. He went on to own part of Aspen and Vail, a meat packing plant . . . he could do anything! He was really a good negotiator, but he knew absolutely nothing about the radio business, and he knew he didn't. I admire a guy who admits he doesn't know something. Anyway, he hired Joe and me to be sort of like consultants for WIXY, which didn't make sense because this guy Gene Taylor was already running the station. There was nothing we could do to save it at that point, because we sold it to him for a big number, knowing he would never be able to achieve what we did. We were supposed to listen to the station and give him ideas and stuff like that. That lasted less than a year. Eight, nine months maybe. They were paying us to be consultants, and they didn't want to take any advice from us. It didn't work, so we just bowed out.

The annual WIXY–May Company Christmas Parade—held on Thanksgiving Day so as not to interfere with early-morning shoppers hoping to get a jump on the holiday rush the next day—boasted yet another top-notch list of talent. The 1972 parade included Canton's O'Jays, riding high from their million-selling "Back Stabbers"; Johnny Nash of "I Can See Clearly Now" fame; the Indians' Buddy Bell; the Cavaliers' Austin Carr; the Crusaders' Gerry Cheevers; WUAB's "Super Host" Marty Sullivan; and WJW's Big Chuck and Hoolihan.

WIXY was still a player in late 1972, but it was nowhere near as strong as in its glory days. Against this backdrop, general manager Gene Taylor announced he would leave the station to explore possible opportunities on the West Coast. His replacement was the legendary Norman Wain, who'd been acting as a consultant for the station. Wain's return sent a warning to competing stations—Wain had a proven track record. In his new position, he would also be president of the Westchester Division of Globetrotter Communications. Back at the helm of WIXY, he had to contend with radio's rapid change to the FM band. (Orders for FM receivers in new cars were up 15 percent over the previous year.)

Music was changing, too. In fact, the first eight songs of the 1972 WIXY Top 100 survey were soft rockers, with Gilbert O'Sullivan's "Alone Again (Naturally)" taking the No. 1 spot. Wain invited Eric Stevens to return to his music programming post. Stevens, who had been off working in the industry, agreed to come back. He tells the story of the time he'd spent "in the industry" before returning: "I had been in New York coproducing the second Brownsville Station album, and I got the call that Norm wanted me to come back to WIXY. I wasn't really too happy with production, because I had a bunch of near misses. I found Alex Bevan at D'Poo's Tool and Die Works. I'd produced the Damnation of Adam Blessing, and we had some chart records. But just because you have chart records doesn't mean you're breaking the bank wide open. The Brownsville Station album came out, and 'Smokin' in the Boys' Room' was the last track on side two. Somebody in Maine played this track, and it went crazy. It sold 1.8 million records."

Veteran newsman John O'Day also joined the 1260 staff that December. WKYC released him, Tom Carson, and Jim Rupert when attorney Nick Mileti bought the station and changed the call letters to WWWE. In the midst of these rebuilding changes, WIXY's ratings continued to slip behind those of sister station WDOK-FM overall and well behind WERE's Gary Dee in morning drive. WIXY had some top-notch talent. Now it needed ratings to match.

1973

The new year brought a flurry of news reports signaling drastic change on the national scene. In January, abortion was legalized, with the Supreme Court's decision in *Roe v. Wade.* Watergate defendants were being found guilty, and Richard Nixon's presidency was crumbling. The Paris peace talks aimed at ending the war in Vietnam progressed frustratingly slowly.

On the Cleveland radio scene, 1973 started with the announcement that another longtime Cleveland favorite would return to the airwaves—*Chickenman!* The "white winged warrior" (featuring the golden tones of former Clevelander Jim Runyon) would be heard as part of Mike Reineri's morning show. There would only be five new shows, with *Chickenman* taking on the *Earth Polluters,* but that blast from the past was still expected to get people tuning in. Plus, Reineri had been featured on WEWS-TV's *Morning Exchange* and won raves for his appearance, assuring him a return visit in the near future.

By February, 1260's door was again revolving, with Mike Kelly taking a job in the Southwest. Native Clevelander Gary Drake, who came to WIXY via WKGN / Knoxville, replaced him. But the new addition making headlines that February was not at WIXY but WWWE (or "3WE"). Larry Morrow had returned to radio, this time in the morning drive, and his 50,000-watt voice would be yet another challenge to WIXY's attempt to regain the throne. 3WE faced more changes when a federal court ordered the station to reinstate three newsmen let go in the Mileti transition. John O'Day, one of these men, said a return would take some soul searching. He was having a great time at WIXY, but the pay would likely be better at WWWE.

Norman Wain had some TV time that March with a film he produced with fellow Clevelander Dennis Glenn. It was titled *This Year in Jerusalem* and looked at issues faced by Russian Jews immigrating to Israel. Former WIXY newsman Fred Griffith hosted the Saturday night presentation on

WEWS, which included an interview with Rabbi Arthur Lelyveld of the Fairmount Temple, who commented on the film. Griffith would go on to a long and distinguished career at WEWS as host of the popular *Morning Exchange* news and talk show, which helped inspire ABC's *Good Morning America.*

Sadly, death claimed a long-standing WIXY team member in March, with the passing of former program director George Brewer in Nashville. May would also bring sad news out of Chicago with the death of veteran Cleveland broadcaster Jim Runyon. He'd been suffering from cancer and was just forty-three years old when he died. The former host of the *Runyon Room* would be remembered for his subtle humor, his velvet voice, and the friendly demeanor that made every listener believe he was talking directly to them and no one else.

WIXY always drew a lot of interest with its year-end surveys, though a midyear list of the Top 100 all-time rock 'n' roll favorites raised some eyebrows. The song "Brandy" by Looking Glass, a one-hit wonder, took the top position, beating Led Zeppelin's "Stairway to Heaven" at No. 2. "Stairway" had never been issued as a commercial single, though WIXY did play it as an album track. Even so, "Brandy" would never have made a list of that type on the FM band.

That spring, jock Frank McHale, who used the name Johnny Michaels when he was on 1260, also returned to Cleveland airwaves for a regular engagement. McHale had since been to WHK, had done plenty of freelance work, and could now be heard on WQAL-FM. Norman Wain saw the obvious trends toward FM, so offered a unique promotion that would help WDOK but not WIXY. He pitched a "Converter Saturday" that offered discount prices at various locations to convert AM radios to FM capability. And he planned to do it regularly in the future as well.

The June ratings book showed WIXY in the top spot among listeners age twelve and over, though Gary Dee remained the king of morning drive, with Mike Reineri at No. 5. WIXY also dropped to No. 6 among listeners eighteen and over. The station was still a good investment but clearly needed work to regain its former dominance, if that was even possible. The owners took the first steps when they brought back Norman Wain and *Chickenman.*

That July, yet another familiar name was on the horizon. WIXY announced that July that Dick "Wilde Childe" Kemp would return to its 6 to 10 P.M. slot. Kemp had been a key player in the station's heyday, had

WIXY's experiment with Dick Kemp as a talk show host was trumpeted from area billboards. (Photo by George Shuba)

always been a free spirit, and when Wain asked him back, he was living above the Colonial Truck Stop in Oberlin, which he and his wife were running, having closed their antique business. Kemp's return to 1260 would seem to have been a bumpy one. He told an interviewer shortly afterward,

> I became unhappy with radio when I went to McKeesport. . . . Personality people were out. The machine types had come in. After working there for eighteen months or so, I had a shot at going to WMMS for progressive rock, and that drove me right up the wall! That put me out in the ball park . . . not because of the music, 'cause I loved it. I couldn't stand the idea that radio was anything other than entertainment. A fun

thing became a bunch of socio-political crap that I didn't want to hear anymore because I heard enough of it all the other times. . . . If I want to take up a campaign in my life, I will take it up and I'll work at that one thing until I finish. In that type of radio [FM progressive rock] they don't. . . . But the music is fantastic. I love the music! A good hard rock 'n' roll boogie situation! But nobody has any time well spent in that type of radio. You're going to be gloomy because you're so drug down by socio-political things.

Kemp was also dissatisfied with the personnel on FM radio; in the same interview, he takes aim at his former coworkers:

They're more plastic [the FM jocks]. I can give you a rundown. I can't give you one guy in this whole town who's sincere in what he does. The rest of them . . . same old thing. Same old rip off. Go in there and screw people over, and I don't like that pseudo-social crap. . . . But WIXY is a happy, fun-filled situation. It's also not hip! It's like a Sunday school picnic if you're having a good time . . . and the money ain't all bad. Throw me some good music, and I'll blow the shit out of it!

Although Kemp did pull in a good chunk of change, he was always known to drive cheap used cars. Bob Bassett, the second disc jockey named Mark Allen to work at WIXY, tells us the reason was obvious: "I asked Dick one time why he drove beaters. He said [it was] because sometimes he would forget where he had parked one after too much partying!"

Kemp's return bumped Terry Stevens to the 2 to 6 P.M. slot. Kemp came out of the box strong, with a format far from the one he'd worked with at WMMS. His first promotion was a contest to meet the Osmond Brothers. Listeners sent in postcards, and he drew as many as a thousand postcards a day, more than fifteen thousand in all. Donny himself was so impressed that he came in to draw the winning entries.

Bob Shannon had had enough by August and left the station, to be replaced by all-nighter Gary Drake. WGCL's Hal King, formerly Helmut Kerling, came in to take Drake's old place. The twenty-four-year-old King made no bones about saying his hobbies included drinking and chasing girls, but he also said, "I'm taking a philosophy course at John Carroll [University] to keep my head in gear. That night shift does strange things to you." King had worked in a brickyard to finance his college career prior to a stint at WNOB-FM, and he said radio work was a lot easier, not to mention a lot of fun.

Billy Bass was certainly a pioneer at WIXY. He was the first African American jock on the station's staff, but he wasn't the last. Also in August, the station welcomed John Adams High alumnus Bill Black, who had also been a personality at WJMO. Black took over the late night 10 P.M. to 2 A.M. shift and was one of the few people of color appealing to a mostly white audience at any station. When asked if he thought he might be a "token black," he quickly, wittily responded, "The thought went through my mind that WIXY was in need of black listeners, so they hired me. But I don't want to chase away Ku Klux Klan listeners just to get a black audience. I don't consider myself on the air just to get black listeners. I'm there to do a job. If I begin to feel like a token black I will sit down with management and tell them that's not where my head is. The job at WIXY is a chance to do a different type of radio."

While WIXY's staff roster constantly changed, promotions remained a constant at the station. "They're Coming to Take Me Away, Ha-Haaa" by Napoleon XIV, which had been yanked from the airwaves because of concerns that it mocked the mentally ill, was rereleased in August 1973, seven years after it originally came out. This time, WIXY linked to it a contest to win a vacation home. But the song barely made a ripple on its second go-round.

Also in August, WIXY announced "the greatest contest in radio history—even bigger than last week's! We probably will never have a contest to top this one until late, late in September." The contest centered on the phrase "WIXY Goes Bananas" and kicked off with songs from the weekly survey, with the word "banana" not-so-subtly dropped into their titles. It included "Delta Banana" instead of "Delta Dawn" by Helen Reddy, "Saturday Night's Alright for Bananas" by Elton John, and "Bad, Bad Banana Brown" by Jim Croce. It could have been a joke or just a typo, but the list also included "Banana Grove" by the "Doodie" Brothers. Bob Bassett called it "a short-lived but fun promotion. Mike Reineri even did a publicity picture, naked except for a bunch of bananas, on the couch in the program director's office. Terry Stevens dressed up like Tarzan and tossed bananas from a covering over the front door at 3940 Euclid to passersby."

WIXY staffers also sold the fruit at the corner of East 9th and Lakeside to benefit the Police Athletic League, staged a Mike Reineri "Banana

Bite-Off" eating contest, and sold a *Top Bananas* LP, with fourteen recent hits, to benefit the Muscular Dystrophy Association. But no one was quite sure what the promotion was really about, so it was quickly retired.

WIXY was clearly not on the most solid ground as it faced a changing music scene, not to mention the growing threat of FM radio. That year's Thanksgiving Parade might have been an indication. In the past it had attracted popular recording and TV stars, but the 1973 version could only secure one headliner, Paul Anka. It also promoted CEI's Reddy Kilowatt, the Sokol Tyrs gymnasts, and Bill Boehm's talented Singing Angels—all of this a far cry from the star-studded parades of years past.

More than the parade itself, Bob Bassett remembers the hours he spent looking for a meal afterward.

> The promotion director, another jock, and myself somehow missed a ride to an after-party. So the three of us roamed downtown on a Sunday, looking for the party or just some food in golf carts we had used in the parade. Later, as we munched on burgers at McDonald's on East 55th Street, the police arrived saying to us "We got a report of some drunk drivers on golf carts. You fellas wouldn't know anything about that would you?" After much stammering and stuttering, one of us explained we had lost our ride and our friends from the station after appearing in the parade. One cop asked "Are you guys deejays?" I said, "Yes." The other cop just said, "Behave," and they left us to our meal and strange looks from the locals.

By the end of the year, Norman Wain found himself in very good company when it was announced that he was on President Richard Nixon's "enemies list," along with Senator Howard Metzenbaum, whom WIXY had endorsed in the 1970 Senate race; insurance executive Peter B. Lewis; and Cleveland Indians CEO Alva "Ted" Bonda, among others. Most were targeted for actively supporting George McGovern during the '72 presidential race. Wain told the *Cleveland Press,* "I'm flattered to be in such prestigious company! I think that protest or an alternate point is democracy, and without it we don't have a true democracy." White House counsel John Dean had turned the list over to the IRS, though it's not certain that anyone who had been named had actually been audited.

That December, morning man Mike Reineri was still posting a strong No. 4 in the ratings, behind WJW's Ed. Fisher, WERE's Gary Dee, and

WGAR's John Lanigan, but WIXY had slipped to No. 5 overall. However, Reineri was confident the station would regain its audience and would be signing another three-year deal. That year, eight of the first ten tunes in the Top 100 were "middle-of-the-road" tunes.

1974

The radio wars raged on into 1974, and Gary Dee, with WERE's People Power format, remained the big threat. And WIXY was dead set on regaining its former glory, despite the new challenges in the medium. Eric Stevens won recognition from the radio industry tip sheet *Pop Music Survey* with his Program Director of the Year nomination. The same publication nominated Marge Bush Music Director of the Year and WIXY Radio Station of the Year.

The March ratings book showed a disturbing trend for AM programmers. WERE was No. 1, overall, followed by WQAL and WDOK, both with FM soft-music formats. WGCL-FM came in at No. 8, ahead of WIXY and WMMS-FM at Nos. 9 and 10. Gary Dee was listed as top dog in morning drive, followed by WJW's Ed. Fisher, WGAR's John Lanigan, WJMO's Rudy Green, and WWWE's new entry in the wars—Larry Morrow. For the first time in a long time, WIXY failed to make the top five in morning drive. Word also came down that WHK would switch to a country format with WIXY vet Joe Finan on mornings.

In an attempt to jump-start his show, Reineri staged an April Fool's Day joke that may (or may not) have backfired. On April 1, 1974, he reportedly phoned in his resignation to station general manager Dick Bremkamp. Not realizing it was a joke, Bremkamp told Reineri this was a contract violation, but said that if that's what he wanted, he was out. By the next day, Bremkamp was waiting to hear on a course of action from WIXY owners Globetrotter Communications, and Reineri was quoted as saying, "I was just trying to have a little fun, but I think I'm stuck now." It's very likely the whole controversy was simply a setup to get press, because Reineri was on the job soon after. He also posed for a (clothed) centerfold poster available around the city. Also in April, Joe Finan joined the newly country-format WHK; the reports were correct!

While Reineri's April Fool's Day gag drew some attention, WIXY's ratings problems couldn't be solved with promotions tricks alone. That month WIXY management apparently branched out in hopes of reaching a broader audience. There was talk on the street that not only was WIXY trying to appeal to more black listeners with more urban sounds, it was also pursuing sales among the African American business community. Indeed, WIXY had been airing a greater number of songs by black artists in a "rock and soul" format. Of course, it's not uncommon for radio salespeople to pursue all possible opportunities.

Also in April, Cleveland suffered a major loss in its broadcast fraternity, with the death of former WIXY morning man Howie Lund. The body of the forty-nine-year-old Lund was found in the bathroom of his Parma home. Coroner Dr. Sam Gerber ruled his death accidental, caused by ingesting Lestoil, a liquid cleaner. Gerber said there was no evidence of alcohol in Lund's system but he had taken a "medicinal" dose of Librium, which may have affected his judgment. Gerber said the amount of Lestoil ingested was "about a mouthful," very little, but enough to have fatal consequences.

On the promotions front, the station decided to pull out all the promotional stops heading into the warmer springtime months and it pulled off some high-profile campaigns. The promise of backstage passes for Grand Funk Railroad drew thousands of postcards. And the Great WIXY Kiss-Off, held at Parmatown and Richmond Malls, brought heavy TV and newspaper coverage. The Kiss-Off was an endurance test to see how long couples could stay with their lips together for fifty-five minutes every hour, with a five-minute rest period. The winners traveled to Chicago for the national contest in an attempt to break the national record of ninety-six hours. Contestants also received T-shirts and albums from the rock group KISS. These kinds of events could draw heavy media interest, but the important thing was translating that interest into ratings—and only the listeners could do that.

Another stunt that May got plenty of publicity, but not the kind the station wanted. Dick "Wilde Childe" Kemp announced he would try to break the world's bull-sitting record by sitting atop a fake bull outside the Cleveland Arena, which was hosting the Longhorn Rodeo. Bob Bassett recalls that the station got the bull from a local Brown Derby restaurant. Listeners were asked to guess when he would dismount and send their answers to the station on a postcard. Winners who guessed closest to the exact time could win an express paid trip to the Longhorn Loretta Lynn Classic Rodeo in Nashville and tickets to the Grand Ole Opry. The promotion started on a Tuesday evening, but soon after, on that same day, someone drove by and fired a pellet gun at Kemp. Pellets are nowhere near as dangerous as larger firearms, though they can still do serious damage; Kemp suffered a slight wound, but the promotion came to a quick end. The station offered a $1,000 reward for information leading to the gunman's arrest.

In summer 1974, WIXY's music mix could be considered confusing at best. 1260 played everything from Ray Stevens and Marvin Hamlisch to the Stylistics and Grand Funk Railroad. Audiences were becoming far more discerning in their musical tastes and lacking the patience to sit through several artists until they got to one that they liked. Program director Eric Stevens left the station in May to once again devote himself full-time to producing records. Stevens had a keen ear for music and trends, and it wasn't likely he would be out of the radio scene long. His replacement was another Stevens, Terry.

Also that month Gary Drake exited, making way for Mike Collins, who left WHK-AM when it went country. Collins's show would prove to be an attention-getter for the 10 A.M. to 2 P.M. audience heavy with female listeners. He did longer talk segments, asking questions such as, "If you were raped, would you report it to police?" and "Would you feel better if your daughter was on the pill?" The majority of the topics he covered were highly controversial for the time. At one point, he even asked women who had had abortions to call in and tell him about it.

In summer 1974, while WIXY was in a constant state of change, music listeners all over the country were enjoying the era of giant rock festivals. The World Series of Rock show at Cleveland Stadium that June featured an all-star lineup, including the Beach Boys, Lynyrd Skynyrd, and Joe Walsh. WMMS-FM cosponsored the Belkin production, but WIXY responded to the show by announcing it would stage another Appreciation Day show that summer also expected to draw as many as 100,000 people.

The spring ratings results came out that June, and it was clear that WIXY needed some tinkering. It had solid programming and personalities, but it faced some very strong competition from aggressive morning men like Gary Dee and John Lanigan, not to mention the increasing switch to FM. The numbers showed the effects of more tuners being installed in cars: four FMers—WQAL, WDOK, WMMS, and WNCR—had moved into the top ten. Gary Dee ranked first in morning, with WGAR's John Lanigan at a strong No. 2, followed by WJW's Ed. Fisher and WWWE's Larry Morrow. WHK's Joe Finan came in at No. 8, pushing Mike Reineri to ninth place, followed by Debbie Ullman at WMMS, at No. 10. Three names linked to WIXY were now competing in the top spots for morning listeners. And WWWE came in second to WERE overall, though 3WE's standings were based to a great degree on Indians baseball, which was traditionally a major draw, despite the team's spotty record.

WIXY wasn't panicking, but it was concerned. The station needed to get its call letters in people's faces. The promotions people had always done a good job planting items in the *Press,* the *Plain Dealer,* and even *Scene,* and there was a seemingly endless stream of contests. That July an internal memo from WIXY promotions person Merrill Colegrove to general manager Dick Bremkamp showed an order for ten thousand "rumper stickers" (at a cost of just $685). Bumper stickers were a low-cost item that could be linked to any number of contests and stayed put once they were affixed to a car, effectively giving the station thousands of rolling billboards around Northeast Ohio.

Improvisational ensemble comedy à la Chicago's Second City was becoming increasingly popular in the mid-1970s. Many artists, like those at the Firesign Theatre and the Conception Corporation, had drawn good response on the FM band, and a troupe working at Pickle Bill's Saloon in the Flats, the Common Sense Novelty Comedy, filled in for a week while Mike Reineri recharged his batteries on vacation. WIXY also added the very popular *Dr. Demento* syndicated show to its Sunday-night lineup. Demento (aka Barret Hansen), who had made a career of airing the most bizarre recordings he could find, proved to be a welcome addition to the weekend programming.

Bill Black had left WIXY in July, and the station named Chuck Baron from WAYS / Charlotte as his replacement. Terry Stevens was quoted as saying Baron would do a "down-home boogie rock 'n' roll show like you never heard before," and if his hype was accurate, Baron would be a player. Word was, he drew 70 percent of the audience on his 6 to 10 P.M. show in North Carolina. There was one problem, though: WAYS wouldn't let Baron out of his contract. Thus, the other jocks had their shifts extended until the station could find someone to take Black's place.

The station finally found its new late-night man: twenty-one-year-old Tim Byrd from WAPE / Jacksonville. Byrd made no bones about declaring

Tim Byrd made it clear that Cleveland was not his first choice for a new radio market. (Photo by George Shuba)

to a *Cleveland Press* reporter that he wasn't all that crazy about relocating to Cleveland. He hadn't heard much good about the city and said the only reason he accepted the WIXY job was that it was a bigger market. He also wasn't impressed with the competition on WGAR, WGCL, and WMMS. Known as the "Birdman" on the air, the new jock would have a big job winning over listeners, especially given his disparaging comments.

The station liked the talk segments in midday but wanted a new voice in that time slot; as a result, it changed several of its jocks' time slots. Mike Collins was shifted to the 10 P.M. to 2 A.M. shift, and Dick Kemp was reassigned to keep the controversy going on the "rock that talks" midday show. He also dropped the "Wilde Childe" from his name that October when he started his new shift. Collins's late-night stay was only temporary; he was switched again, to weekend duty, when the new late-night jock, Chuck "Bobo" Baron, arrived from WAYS following the resolution of his contract problems. Tim Byrd would now take Kemp's place from 6 to 10 in the evening. It looked good on paper, but the revolving door spun a lot faster than usual that month. Dick Bremkamp left the station, and, after just one day on the job, Baron decided to leave because of the change in management. Mike Collins stayed put, at least for the time being. Bremkamp had reportedly parted ways with WIXY over "a conflict in programming philosophy." His replacement was Gunnar Bennett from the Globetrotter Communications Chicago offices. Meanwhile, WJW radio quickly signed on Bremkamp.

WIXY did its Top 100 of 1974 countdown on New Year's Eve, and this year's list was heavy on middle-of-the-road music. The No. 1 song of the year was "Billy, Don't Be a Hero" by Bo Donaldson and the Heywoods, followed by songs by Terry Jacks ("Seasons in the Sun"), Olivia Newton-John ("I Honestly Love You"), and Paul Anka ("You're Having My Baby").

1975

The station started 1975 with a new midday man, Tom Kent, replacing Terry Stevens. Kent was a true child of the media. "I grew up in Winston-Salem, North Carolina," he said, "and caught the radio bug back when personality was king. I used to hear these great jocks on the air and just imagined that would be the dream job. I started my own radio station out of my bedroom, complete with transmitter, and would get my first job on the air at fifteen at Top 40 station WAIR." The job at WIXY was an obvious career move for Kent, so much so that he would pass up an audience with the King for it. He recalls, "I was working at WHBQ in Memphis. I was only nineteen years old. I was doing this crazy 'Truckin' Tom Cookin' Kent' high-energy gig there when one night my program director, George Klein, came into the studio to tell me some startling news. You see, George was Elvis's best friend. He and Elvis went to Humes High School together and remained very close friends over the years. There George was, in the studio while I was on the air, to say that he was just up at the mansion and that him and Elvis were listening to me and Elvis wanted to meet me. He set it up for me to meet Elvis at the Memphian Movie Theater in downtown Memphis. Elvis would rent out the theater and show old movies all night and you would go and hang out with Elvis. It was all set up for me to meet the King when I got the call to go do nights at WIXY 1260. I was single, and radio meant everything to me, so I took the job in Cleveland and never got to meet Elvis." Even so, he was hesitant about moving north: "I was only nineteen, had never lived up north before, and had a lot of reservations about coming to Cleveland. It was a culture shock, and it was damned cold!"

Paxton Mills joined the lineup as well. He came in from KLIF / Dallas and signed on to do the 2 to 6 P.M. show. Mills and Kent hit it off, but it wasn't so easy with the rest of the staff. "I was the youngest person at

WIXY was such an important career move that Tom Kent gave up a meeting with Elvis Presley to start at the station as soon as possible. (Photo by George Shuba)

every station I worked at," Kent said, "and it always created a little bit of a bias with the older guys. At WIXY, people like Mike Reineri sort of treated me like this punk kid. I remember Paxton Mills kind of treating me like his kid brother. It was all good though. There was a perception, I'm sure, that I hadn't paid my dues."

Kent strongly remembers many small details about other members of the WIXY team. For example, he says, "I used to hang out with Tim Byrd, the 'Birdman.' We became very good friends and still are. Tim had a very weak constitution, or perhaps a very consistent one. It seemed like every day he played 'Freebird' by Lynyrd Skynyrd at the very same

Paxton Mills
brought a new
sensibility to
WIXY when he
took over after-
noon drive du-
ties. (Photo by
George Shuba)

time. I would come into the station and no one would be in the studio
and I would find Tim in the can doing his thing."

Byrd and Mills also put Kent through a sort of on-air hazing before he
could claim his stripes. "I remember my first night on the air and Paxton
Mills and Birdman setting me up to fall on my face when I first went on.
All the music was on cart, and they had me set up so that the cart would
fail when I first pushed the button. I was all hyped up and ready to go,
and I'm hitting the button and nothing's happening. I can hear them both
in my headphones in the background laughing. These were the kind of
pranks you had to endure when it was your first time and when you were
only nineteen."

Paxton Mills also brought a new dynamic to the station: he could sense
club hits that would make it on the air. The disco craze was taking root,
and some songs were hits in clubs and dance halls months before you

heard them on the radio. In truth, the AM musical landscape was start-ing to wear very thin, with a top-rated station like WIXY playing songs like Barry Manilow's "Mandy" along with lighter fare from Grand Funk, like "Some Kind of Wonderful." Even so, it still offered good opportuni-ties to see talent on stage for inexpensive ticket prices. WIXY sponsored shows by artists such as Neil Sedaka, Melissa Manchester, Frankie Valli and the Four Seasons, among others, for prices ranging from $4.50 to $6.50, top ticket prices when the Beatles first hit U.S. shores.

The station was obviously struggling, falling to No. 10 overall in the ratings released that March, with Reineri climbing to No. 8 in the morn-ing. WIXY's promotional materials claimed that one out of five young adult listeners in Cleveland tuned in to Reineri. The station rushed out additional material, too, which claimed its male numbers were up 73 percent in morning drive, 290 percent from 10 A.M. to 3 P.M., 119 percent from 3 to 7 P.M., and 143 percent overnight. There were similar increases in female listenership, again thanks to creative interpretation of the rat-ings. The station was now selling twelve commercial minutes per hour, with a boast: "WIXY merchandises, and does it better than most stations across the country." The station also stated, "Nearly one-third of Cleve-land's young adults (ages twelve to thirty-four) listen to WIXY—and people under thirty-five make up almost half of Cleveland." Salespeople stressed the music as "hit-oriented and very familiar" and said the station could "be listened to for hours at a time without becoming repetitious." They also claimed WIXY played more music than its competitors, with personalities who are "real people speaking in a real manner."

Still, rumors started flying about the possible sale of the station along with WDOK-FM to Combined Communications, a group out of Phoenix, which was said to be discussing a switch to a news/talk format with NBC at the top of the hour. By April, the weekly WIXY survey in the *Press* took up fewer column inches and included fewer songs and was joined by the M105 (WWWM-FM) album survey. The new station, with former WIXY wunderkind Eric Stevens at the helm, had signed on to challenge the FM rockers.

But 1260 hadn't given up the fight yet. Promotional material from that month shows WIXY crunching the numbers and stating the Pulse survey showed it as No. 1 with men ages eighteen to forty-nine from 6 A.M. to 7 P.M. and No. 1 with twenty-five-year-old to forty-nine-year-old men 10 A.M. to 3 P.M. The sales department urged clients, "Buy it now, and buy it big!"

Randy Robins joined the staff to do the 4 to 8 P.M. shift, telling the *Press,* "I'm still representing myself, even though I work for WIXY. I have me to sell. I look out for Randy. Everyone should look out for themselves. You are your own product." Those were surprising words from a jock hired at a station known for its team philosophy.

Another rumor making the rounds that spring had WIXY considering an urban format. Globetrotter Communications had a successful urban station in Chicago, though WIXY was still making plenty of money. By that May, talk about the sale of WIXY and WDOK to Combined Communications was starting to fade, though the WIXY School of Broadcast Technique was sold to a group called Educational Broadcast Services.

That June, the station hired Gregg "the Groover" Cleveland. The former Gregg Crawford had most recently worked at KIKX / Tucson, and rejected a move by WIXY to name him Jack Daniels. He was excited about working in a market the size of Cleveland and would be manning the 10 P.M. to 2 A.M. shift.

In June's spring ratings book put four FM stations in the top ten, led by easy-listening WQAL and WDOK. WIXY was completely out of the running, though Mike Reineri managed to squeak in at No. 10 in mornings. Flagging ratings and the uncertainty of the station's format pending a sale could not have made WIXY a very happy place to be. Also, work at WIXY was more than a full-time proposition: records show that, surprisingly, the whole staff, including Reineri and all the other jocks, filled out time cards, listing forty-six hours for a six-day week.

In July, a big change came in the critical morning drive slot. Tom Murphy came in from WMAQ / Chicago to replace Mike Reineri—who left to pursue other opportunities. Calling himself the "World Famous Tom Murphy," he moved to Cleveland because of its reputation as a top radio market. It was now his job to help revive WIXY as a top ratings winner. One of his first stunts was to broadcast from a phone booth, giving out on-air clues to his whereabouts and awarding a new bike to the first person who located him.

The following month, WIXY named a new news director, Chuck Bolland from WKRC / Cincinnati, who would replace Bob Engle. Bolland also had a syndicated sports program, *That's the Way the Ball Bounces.* Longtime voice Bob Engle continued to do news on the afternoon shift.

WIXY tried to one-up the competition with innovative on-air programming, though some of it might seem comical by today's standards.

Some of it was intentional, like the "Superjaws" comedy song the jocks recorded, though others seemed a little odd, like Grandpa Hickabucket's one-man band and the promotions department handing out house plants or "How's Your Love Life?" T-shirts. Tom Murphy announced his write-in candidacy for Cleveland mayor, and during the "Save the *Cod*" radiothon, Paxton Mills lived on the submarine WIXY was trying to save. The station also solicited tennis shoes for a Bruce Springsteen *Born to Run* contest.

And in downtown Cleveland a mysterious masked man in black tights and cape passed out clues to the contents of a box suspended from a crane at the Parmatown Mall. But that contest didn't go as smoothly as everyone hoped. Looking back, Mark Allen (Bob Bassett) says that it seemed to drag on forever.

> Every weekday, more prizes were added, and we would call it "Prize Package Number 185" or some other number, to make it seem like hundreds of prizes. Truthfully, there were many prizes worth lots of bucks. They were mostly trade-outs with sponsors, but good ones, like trips. To win, you had to guess what was in a large shipping crate that was bigger than a car. This crate was moved each week to a different shopping center in the area. A corner near the most traffic was selected for this container to be hoisted high above the ground by a crane.

But that contest had strict rules, and, Allen says, that's where the problems arose. "You could not win if you had won anything in the last month. This fact was stressed in print and on the air. The first person to guess correctly what was in this container [Tom Jones's bow tie] turned out to be a professional contest winner who had lied to me about not winning anything from us within a month. There was a lawsuit; she lost, and we looked stupid getting another question and winner."

November brought the annual May Company Christmas Parade, which the station claimed would draw a quarter million people to the city. The star power included the Michael Stanley Band, TV's Banana Splits, singer Al Martino, and a group that had seen better days but was hoping to revive its career with WIXY exposure—the Jackson 5, with a now teenage Michael still the front man.

Larry "J. B." Robinson had become as recognized as the WIXY jocks, thanks to his jewelry store chain's heavy advertising schedule. He also came to the aid of the station's promotions and sales departments in their time of need that December, with a "Great Diamond Dig" at three mall locations, with contest finalists turning up dirt to find $1,600 worth of merchandise (of course, he benefitted promotionally as well). There was also a Christmas show for the kids at Cleveland Music Hall, starring Bob Keeshan as Captain Kangaroo. Keeshan received the key to the city, and December 22, the day of his show, was designated "Captain Kangaroo Day" in Cleveland. The final determination of those promotions' success, of course, would come with the fall ratings book, released in just a few weeks. No matter how hard the staff worked, it was difficult to be optimistic for WIXY in the current radio scene.

1976

Radio was well into a decade of change, and change at specific stations was often dictated by the ratings. The fall 1975 ratings book was released in the first week of 1976, and the news wasn't good for WIXY. Morning man Tom Murphy failed to make the top ten, and so did the station overall. Plus, the FM stations were showing more strength, with WQAL and WDOK topping the list, and WMMS and the new WWWM (M105) all placing in the top ten. In an attempt to reestablish its music image, WIXY announced it would air four hours of commercial-free music every day. Program director Steve Kelly also promised to play more album cuts and "progressive" music, especially at night. He told *Press* columnist Bruno Bornino, "Our top requests lately have been for 'Love Is the Drug' by Roxy Music, 'Dream On' by Aerosmith and 'Dream Weaver' by Gary Wright. None of these are what you would classify as 'singles' acts, but if that's the music our listeners want, that's exactly what they're going to get." He also promised more "progressive soul sounds," like "Love Rollercoaster" by the Ohio Players, and "You Sexy Thing" by Hot Chocolate.

"But listeners will really notice the changes at night," Kelly stressed. "We'll be playing album tracks from groups like Foghat, Queen, Bruce Springsteen, Bob Dylan, Joni Mitchell, and the Cate Brothers." He also said, "We've been thinking about changing our format for a year now. The straight 'Top 40' isn't making it anywhere in the country today. Young people are just too musically sophisticated. They're not interested in purchasing singles anymore, they're into albums." But the problem at hand went beyond the music selection, as Kelley admitted: "We realize we'll be fighting an image problem at first, but we're optimistic that our listeners will like the change. And we expect to attract many new listeners."

Reading between the lines, it would appear that this was WIXY's last gasp. First, listeners hoping to hear so-called progressive acts weren't

Program director
Steve Kelly had the
unenviable position
of battling the rising
competition from
FM rock stations.
(Photo by George
Shuba)

going to sit through Paul Anka, Al Martino, or the Carpenters. Second,
WIXY's huge image as a teen rocker from the past would work against it.
And, finally, the bottom line was the station broadcast in mono. You had
one-stop shopping in stereo on the FM dial. Ultimately, Kelly's gamble
to include more album cuts didn't pan out, and he was gone by the end
of February, replaced by Bill Bailey from WDRQ / Detroit, who had also
been at WLS / Chicago, and took over the ten-to-noon shift.

Still, one high school class project offered a faint glimmer of hope. A
survey of three hundred students at Normandy High in Parma conducted
by the school's TV Production and Broadcast Workshop ranked WIXY as
the most popular AM station with students, followed by WGAR. The FM
favorites were WMMS, WWWM, WGCL, and WLYT, which had started
playing a format heavy on disco music. However, the survey results did
not mention what little variety there was on the AM dial; when there's
only one hamburger restaurant in town, that's usually where everyone

buys their hamburgers. The poll also showed the favorite personalities all to be FM broadcasters.

Taking a cue from the competition on the other dial, WIXY started a ride service, a way for people needing and offering transportation to exchange information, similar to the People's Want Ads previously heard on WNCR-FM. It also offered listeners a chance to sit in on weekly music meetings to determine which songs the station would play. WIXY improved its signal as well, giving listeners stronger reception at night. Now those potential listeners just needed to tune in.

Trying to capitalize on any trend that might bring in listeners, the station tried to connect itself to the current citizens-band radio craze. The truck-driving anthem "Convoy" by C. W. McCall got heavy play on WIXY, and on the weekend of February 28 the station staged the first annual Citizens Band Radio Fair at the Sheraton-Cleveland Hotel. Sales manager David Ross, very confident about the future of that medium, produced it. "It's predicted that CB sales could hit $1 billion in 1976," he claimed. "They could surpass television set sales by 1980." The station issued convoy code stickers with CD jargon as well. Of course, the CB fad would drop through the basement as quickly as it had risen, leaving WIXY without any long-term ratings boost.

That April, an ad campaign claimed, "There's a WIXY you know. And a WIXY you don't." It showed four pictures of the same person changing over the years. In 1965, he had a clean-cut appearance, and the big things in life for him were "Burgers, Beach Boys, Beatles, Bubble Gum and WIXY." The next photo showed the peace-symbol-clad model and read, "1968. Nehrus, Miniskirts, Monkees, Sonny & Cher and WIXY." 1972's was a long-haired version flashing the two-fingered peace sign and thinking about "Peace, Grease, Police, Moody Blues, and WIXY," and finally, in the 1976 segment, the same guy with shorter hair showed "WIXY's back and better than ever." The ad also promised, "Now we've gone through our changes. Our head's back on straight again. And we're giving you music like you like it, like never before." It also declared, "We're not getting older. We're getting better." The word "older" likely was a red flag for younger people who might have considered tuning in the station, as unlikely as that seemed in the first place. Still, the station pegged its hopes for the future on the catchphrase "WIXY's Back." It hoped the listeners would be as well.

WIXY's final great promotion took place that May and centered on rising star John Travolta, from ABC-TV's *Welcome Back, Kotter.* Travolta had

Future superstar John Travolta thanks WIXY's Marge Bush for supporting his first single, "Let Her In." (Photo by Janet Macoska)

also begun a recording career, and WIXY was giving heavy airplay to his "Let Her In." The station offered a dinner and private party for Travolta and contest winners as well as an autograph party for the public. While in town he also put his feet in cement outside the new Peaches record store (at Pearl and Brookpark Roads) and stopped by the Great Lakes Mall. The appearances gave listeners a chance to see a superstar in the making, prior to his breakthrough performances in the films *Saturday Night Fever* and *Urban Cowboy*. Marge Bush recognized his star potential, recalling, "When Travolta was coming in, *Welcome Back, Kotter* was so big and I told Bill Bailey, the [program director] at the time, to be sure we had enough security. When he laughed, I told him how big the show was, and we'd have a lot of groupies and real fans at the station. Laugh, laugh. The day came, and it was like a madhouse . . . , and we could hardly get in. [Bailey] immediately called for backup security both at the station and the music store where he was going to make an appearance."

Then, after this last wild promotion, the results of the spring ratings book were released, and the door was ready to be slammed shut on 1260. The June book showed Gary Dee at No. 1 in mornings, followed by John Lanigan and Ed. Fisher. WIXY didn't make the cut. As for overall numbers, 6 A.M. to midnight, WWWE-AM (a sports-heavy station no doubt helped by the red-hot "Miracle at Richfield" Cavaliers) took the top position, and, again, WIXY didn't even make the top ten.

Wheels were now in motion for drastic changes. On Monday July 19, 1976, WIXY officially changed format to one aimed at the twenty-five-to thirty-four-year-old crowd, though it would likely appeal to a much older demographic. Globetrotter Communications filed an application with the Federal Communications Commission to change the call letters to WMGC-AM, the call letters suggesting "magic radio." The artists now being played included the Carpenters, Cher, Barbra Streisand, and Neil Diamond. Globe Broadcasting promised three or four songs in a row, often aimed at setting a particular mood. For example, if Cat Stevens sang about the problems of everyday life in "Wild World," it would be answered by James Taylor's "You've Got a Friend." The station's slogan was also changed to "The Music Is the Magic."

Change of this type on the air often means a transition with the staff, and that was the case with "magic radio." Program director Bill Bailey, who had also been doing afternoon drive, turned in his resignation. Tom Murphy left morning drive a week later for a gig at KGIL / Los Angeles, and Paxton Mills simply walked down the street to do morning duties at WWWM-FM.

There was also a persistent rumor that "magic radio" might go through another transformation in the very near future. Talk centered on NBC approaching the station to carry its News and Information Service as soon as WERE, which now had an all-news format, dropped it. The management at Globe actually considered this all-news route for a time but ultimately rejected it. A new lineup was in place by mid-August: Rick Monroe in mornings, Phil Thomas from noon to 6 P.M., Brother John Letz from 6 P.M. until 12 A.M., and Tom Smith taking over until 6 A.M.

But WIXY wasn't leaving the headlines yet. Just a month later, the Cleveland chapter of the National Organization for Women asked the FCC to investigate charges of bias against three VHF television stations and seven radio outlets, and WIXY was among them. NOW charged them all with sex and race discrimination for failing to hire more women, blacks, and Hispanics. The powers at 1260 quickly moved to remedy the situation.

The station's call letters officially became WMGC in the second week of September. It also brought back a big-name player from Cleveland's rich radio history: Jeff Baxter, formerly part of "Baxter and Riley" at WERE, and brought back Jack Riley for a weeklong reunion in October. Meanwhile, Brother John Letz barely lasted a month before being replaced by Christy Phillips, who would become the first female afternoon drive disc jockey on Cleveland AM radio. She'd been a fill-in jock and news reporter at WIXY and would say she was at the right place at the right time when this job became available. A steady stream of personalities passed through the doors at WMGC during its short history. Few listeners remember WMGC today, save for its controversy over a billboard that included the slogan "Get Your Rock Soft." On April 14, 1979, WMGC changed to a talk format, with the call letters WBBG-AM, which is said to have stood for "Boys from Bowling Green," for the owners' alma mater.

EPILOGUE

Paul McCartney commented on his ups and downs with the Beatles by likening the group to a vacation where it rains six days out of seven: you would always remember how great that one dry day was. That could very well describe how many of us feel about WIXY 1260. At its prime, the station seemed invincible—taking on competitors with seasoned performers and stronger signals and leaving them with no choice but to change formats if they hoped to survive. WIXY's jingles, air sound, music mix, and especially its jocks had become part of our personal histories, and the memories grow fonder with each passing year. Joe Zingale smiles when the call letters are mentioned: "Yes, it is amazing how many people remember it and smile as they say it," he recalls. "Hardly a car drove down the street without WIXY blaring out of the radio."

Memories are important, and, as Bob Weiss points out, they can take us to a happier, youthful place. "You always remember your firsts," he says. "Your first car, your first love and, in some cases, your first radio station." Often, WIXY was a part of all three.

Norman Wain adds,

Bob Weiss, Joe Zingale and myself . . . Eric Stevens, Bill Sherard. We've all had the same conversation. When we were going through it we had no idea we were pioneering or doing something different. We were just guys trying to make a buck. It was a business. I got up every day and went to work. Just like the original Beatles: nobody knew that they were going to be with us the rest of our lives. It was surprising to us that people got so emotional about it.

And people did, indeed, get emotional about it. Even many years after the station closed down, Clevelanders share WIXY stories with Wain. He remembers:

Many years ago I had an office on Chagrin Boulevard. I was talking to a woman about some other subject. I forget what it was. Then she asked, "Are you Norman Wain from WIXY?" "Yeah!" I happened to have some tapes of the WIXY jingles. I played them and she burst into tears. I said, "What happened?" "Oh," she said, "My whole life came back to me. That's when I met my husband." . . . There's an emotional connection that we had no idea we were making. I recently met up with Sister Mary Ann Flannery at John Carroll University. She's the head of the communications department there. It turns out she was the principal of Lumen Cordium High School in Bedford at the time we were running High School Spirit. She remembers the kids sitting in the lunch room writing their names to send in the petitions. Sister Flannery told me, "We won the juke box!" Even to this educated woman, religious and all that stuff, it was part of a growing up period. To us it was the way we made our living.

Weiss, Wain, and Zingale, and their families, are all still close friends. They still get together every year, usually in Florida, to look back and even to wonder what might have happened if they had continued. Weiss imagines, "We might have taken WIXY to FM. Stations like WGCL [also in Cleveland] worked. We would have likely created a dynasty, one of the biggest broadcasting groups ever." With their track record there's no doubt that prediction would likely have come true.

Bill Sherard says he's not surprised at the continued interest either; he calls WIXY "one of America's premier stations that created an image built on creativity, excitement and constantly developing variations on a theme. There are not many stations in the country that were as creative. It's a short list."

A lot of people learned lessons that would follow them through their post-WIXY lives. Candace Forest remembers the inventive, wild days at WIXY, especially for the education it gave her:

WIXY made [TV's] WKRP look like a "normal" day. We were creative and crazy, and we went from the absolute bottom of the rung all the way to the top. We were a team. George Brewer was my best friend. He was brilliant and witty and a great radio man. We were part of the community. We responded to the people, and we were always doing unexpected, adventurous things. There's just none of that anymore in commercial radio. . . . Radio now is almost unlistenable. It is boring and without character and without excitement and without creativity and without fun. We were breaking new ground at WIXY all the time. I was very lucky to be part of it. It was a hard job, but it was a great job, . . . I am still using every single skill I ever learned there in my life today. Norm Wain was a great teacher. He knew more about promotion than ten people, and I will always be grateful to him for giving me a break. I was one of the first women in probably the whole country to have that job. After me, there were lots. It was exciting to break that ground, and I am proud of what we accomplished.

When the end seemed imminent, Bob Bassett chose to leave, explaining: "I am, and always will be, a radio fan first. I did not want to turn out the lights at the greatest station I ever worked at." He adds,

WIXY will be remembered fondly by thousands of boomers and others who remember cruising in our first car with an AM only radio, all the great remotes, promotions, stunts, celebrities, and jocks who did things that had not been done before. WIXY had a crazy bunch of promotion folk who were unafraid to risk our lives. Due to this, the bond between the listeners and us was and is strong. WIXY was their station. As the listeners slowly faded into FM, they never saw or heard anything from us that suggested it would ever end. No one turned out the lights in our hearts.

Chuck Knapp also beamed with pride on a visit to Cleveland years later, when he saw an artifact that brought back those times. "When I visited the Rock and Roll Hall of Fame, the radio showcase had a bright blue WIXY sweatshirt with a red S on the front. Our names were on the back. WIXY loved the listeners, and that's how I'd like to have it remembered."

Looking at WIXY and comparing it, and its time, with the stations in today's industry, Tom Kent says,

I think it will be remembered as one of the all time great radio stations in the personality era of Top 40 radio, which, I want to say, spans from 1964 to 1984. There's a good twenty years there when radio was pretty darn happening. The question becomes, "Twenty years from now, will anyone care?" Radio is dying, because it has become all about the money. Since deregulation and the consolidation and corporatization of radio, it has systematically been diminishing . . . to the point where this next generation of radio users don't really care. They have been programmed not to care.

Eric Stevens can say he once programmed two stations that would have the "magic" name associated with them. WIXY became WMGC, and WWWM, better known as M105, would become WMJI, or "Majic."

In the 1980s, the station hosted WIXY reunions, bringing in the former staff to do air shifts and share memories with the public about a time most held as dearly as their listeners did. Norman Wain tells his story of the reunions, and what they did for the WIXY mystique, very clearly: "One of the things that sustained the WIXY legend was WMJI with the basement tapes. They claim they went down in the basement of the old WIXY building and they found these tapes, and they used to have these WIXY weekends. Of course, people remembered it, and they loved the jingles, and sometimes the jocks would come back. King Kirby would come up from Mexico and do an hour or two and it sustained the aura."

Lou "King" Kirby, Jim LaBarbara, Billy Bass, Chuck Dunaway, Norman Wain, Paxton Mills, Chuck Knapp, Jackson Armstrong, and Larry Morrow all joined in to recapture WIXY's heyday, and the reunions were a tremendous success. For a weekend at a time, listeners could travel back to an era when, despite all the hardships of a complicated world, they could escape with an AM radio and know better things had to be ahead. For that alone, we should all tip our hats to a very talented bunch of people, known on and off the air as the WIXY Supermen.

BIBLIOGRAPHY

ANONYMOUS

"'Big Ralph' Show Replaces 'Big Jack' on WKYC-TV." *Billboard,* September 16, 1967.

"Billy Bass Leaving: Three Changes Stir WIXY Radio Staff." *Cleveland Plain Dealer,* December 17, 1970.

"City Needs Queen for a Birthday." *Cleveland Press,* March 19, 1971.

"Cleveland Hot 100 Outlets Go Literary in Promotions." *Billboard,* October 28, 1967.

"Format Change Keys R & B Nights at WKYC." *Billboard,* January 27, 1968.

"Most Here Say It's Great to Be on Nixon List." *Cleveland Press,* December 21, 1973.

"Norman Wain to Return Jan. 2 to Manage WIXY and WDOK." *Cleveland Plain Dealer,* December 12, 1972.

"Playmate Plays Ball." *Cleveland Plain Dealer,* April 12, 1970.

"Two Tie Francine in Tape-Measure Finish." *Cleveland Press,* September 25, 1968.

"Wall Street Panics Over Francine's Figures." *Cleveland Press,* September 20, 1968.

"WIXY Gets New Manager, Replacement for Pixie Trio." *Cleveland Press,* November 10, 1971.

"WIXY Sues Strikers for $500,000." *Cleveland Plain Dealer,* June 8, 1966.

"WIXY's Triple Play Wins Ratings Game." *Billboard,* November 5, 1966.

"WIXY's Top 60 Chosen After Careful Study." *Cleveland Press,* January 12, 1968.

"WKYC Bows Power Radio with a Capital Promotion." *Billboard,* February 10, 1968.

"WKYC Takes No. 1 Spot in Cleveland." *Billboard,* February 5, 1966.

BYLINED ARTICLES

Barrett, Bill. "Disc Jockeys Are Nice Guys: Vote for Your Favorite." *Cleveland Press,* November 11, 1970.

———. "Docile Isn't the Word for Our Jack Armstrong." *Cleveland Press,* October 7, 1966.

———. "50,000 Watts and a Contract Beats the WIXY Pixie." *Cleveland Press,* January 11, 1967.

———. "Finan Won't Be Back." *Cleveland Press,* January 27, 1966.

———. "Howie Lund Left Jobless Over a Matter of Principle." *Cleveland Press,* June 14, 1966.

———. "New Beatles Price Scale: Generosity or an Omen?" *Cleveland Press,* May 5, 1966.

———. "Poet at WIXY Radio." *Cleveland Press,* September 9, 1969.

———. "Readers Write and Writhe about Program Changes." *Cleveland Press,* January 10, 1966.

———. "Takes Radio Station to Task for Promoting 'Sickly' Tune." *Cleveland Press,* July 26, 1966.

———. "The People Write, and Many Are Plenty Mad at WIXY." *Cleveland Press,* December 20, 1965.

———. "Want to Feel Like a Pixie? Try New Sound of WIXY." *Cleveland Press,* December 8, 1965.

———. "WIXY and WJW Close the Rock-Music Gap." *Cleveland Press,* December 21, 1970.

———. "WIXY Balloon Followers Trample Lawyer's Garden." *Cleveland Press,* July 9, 1968.

Bornino, Bruno. "TV's Barbarino to Receive Welcome Here May 3." *Cleveland Press,* April 23, 1976.

———. "WIXY to Lose Teen Image by 'Magic' Monday." *Cleveland Press,* July 16, 1976.

Brewer, George. "WIXY's Brewer: Back to Basics." *Billboard,* August 31, 1968.

Eszterhas, Joseph. "'43-24-37': 'Francines' Here Jam City's 'Wall Street.'" *Cleveland Plain Dealer,* September 26, 1968.

Harig, Judith. "Radio Show Ponders Teen-Ager's Problems." *Cleveland Press,* April 16, 1962.

Hart, Raymond. "WIXY Radio, Globetrotter Reach Merger Agreement." *Cleveland Plain Dealer,* April 24, 1971.

Peters, Harriet. "At White WIXY: Black of Night." *Cleveland Press,* February 1, 1974.

———. "At WIXY: Morrow Today, Shannon Tomorrow." *Cleveland Press,* June 30, 1972.

———. "On Night Shift, His Head Is out of Gear." *Cleveland Press,* March 15, 1974.

———. "Untimid Tim Finds Cleveland for the Byrds." *Cleveland Press,* September 13, 1974.

Reesing, Bert. "Lovers of Good Music Protest WIXY's Switch to Rock 'n' Roll." *Cleveland Plain Dealer,* December 15, 1965.

Stock, Robert T. "Yells in Square Pick Miss Hot Pants '71." *Cleveland Plain Dealer,* May 8, 1971.

INTERVIEWS

Jack Armstrong, via e-mail. Spring 2005.
Billy Bass, Cleveland, Ohio. January 2005.
Bob Bassett, Cleveland, Ohio. Winter 2006.
Marge Bush, Brecksville, Ohio. Autumn 2005.
Chuck Dunaway, via e-mail. Summer 2005.
Candace Forest, via e-mail. February 2006.
Al Kazlaukas, Cleveland, Ohio. Winter 2006.
Dick Kemp, recorded on tape by unknown interviewer. Cleveland, Ohio. 1974.
Tom Kent, via e-mail. Autumn 2005.
Chuck Knapp, via e-mail. Summer 2005
Jim LaBarbara, via e-mail. Summer 2005.
Frank McHale, via phone. Cleveland, Ohio. February 1974.
Linda Scott, via e-mail. February 2006.
Bill Sherard, via e-mail. January 2006.
Eric Stevens, Moreland Hills, Ohio. Autumn 2005.
Norman Wain, Shaker Heights, Ohio. Autumn 2005.
Bob Weiss, via phone. Chapel Hill, North Carolina. January 2006.
Joe Zingale, Chagrin Falls, Ohio. January 2006.

RADIO SHOWS

WWWE-AM, Big Jack Armstrong, "Radio 11 Reunion," 1983.
WIXY-AM, Chuck Dunaway, January 2, 1969.
WWWE-AM, Larry Morrow, November 7, 1981.

INDEX

MIKE OLSZEWSKI is a veteran radio and television personality, historian, and educator. He is best known for his work at WMMS-FM and has written several books concerning the history of Northeast Ohio broadcasting. Along with many regional and national broadcasting awards, Mike won a 2009 Emmy for his TV documentary *Radio Daze: Cleveland's FM Air Wars*. Along with his broadcasting career, he also teaches media and communication courses at Kent State University, Notre Dame College, and the University of Akron. Mike and his wife, Janice, live in Ohio.

RICHARD BERG is a longtime media historian and is recognized as one of the leading authorities on Northeast Ohio radio. His firsthand knowledge comes from close relationships he has established with some of the biggest names in the industry. He is currently working on extensive projects concerning the history of Akron radio and Cleveland's KYW/WKYC-AM.

CARLO WOLFF, a journalist and pop culture historian living in a Cleveland, Ohio, suburb, is the author of *Cleveland Rock & Roll Memories: True and Tall Tales of the Glory Days, Told by Musicians, DJs, Promoters, and Fans Who Made the Scene in the '60s, '70s, and '80s*.